CSR, Sustainability, Ethics & Governance

More information about this series at
http://www.springer.com/series/11565

Philipp Aerni

The Sustainable Provision of Environmental Services

From Regulation to Innovation

Springer

Philipp Aerni
Center for Corporate Responsibility
and Sustainability (CCRS)
University of Zurich
Zurich
Switzerland

ISSN 2196-7075 ISSN 2196-7083 (electronic)
CSR, Sustainability, Ethics & Governance
ISBN 978-3-319-19344-1 ISBN 978-3-319-19345-8 (eBook)
DOI 10.1007/978-3-319-19345-8

Library of Congress Control Number: 2015945024

Springer Cham Heidelberg New York Dordrecht London

Printed on acid-free paper

Springer International Publishing AG Switzerland is part of Springer Science+Business Media (www.springer.com)

Acknowledgements

This book started as a series of working papers drafted during my time at the Food and Agriculture Organization of the United Nations (FAO) in Rome, Italy, from May 2012 to September 2013. At that time, I was coordinating a project titled 'Remuneration of Positive Externalities/Payments for Ecosystem Services (RPE/PES) in the Food and Agricultural Sectors'.

While I was based in Rome, I continued to obtain my salary from ETH Zurich, which was asked to by the Swiss Federal Office of Agriculture (FOAG), the main donor of the project, to head the project coordination team under the guidance of an FAO Steering Committee and the supervision of the former FAO Natural Resources Management and Environment Department.

I started the project by having several bilateral meetings with FAO-based experts on PES. I was quite familiar with the RPE/PES system in Switzerland, which was based on the concept of multifunctional agriculture in which the government provides farmers with generous subsidies in return for providing positive externalities to society and the environment—assuming that markets would not be willing to remunerate such externalities. However, I was not sure how such a concept could possibly work in less affluent tropical countries. When I heard from many FAO experts that there is not much evidence that PES actually does work in developing countries, I decided to visit a developing country where such PES projects have been introduced on a large and small scale by various government and non-government institutions. I selected Kenya because it is widely considered to be a pioneer in the promotion of incentives for the provision of environmental services in agriculture. The field experience as well as the numerous interviews in Kenya confirmed my view that positive externalities in these countries are largely generated as by-products of local entrepreneurship and innovation. Whereas most of the PES projects I visited lacked a concrete business case because they rely almost entirely on external funding, such projects bring together local actors who have not previously interacted. These actors were originally assigned as buyers and sellers of environmental services in the project, but they saw few business opportunities in the rather passive role they were supposed to play. So, many of the actors became entrepreneurs on their own behalf and started to create their own bilateral local business deals for the sustainable provision of environmental services and their innovative practices eventually led to new local markets for environmental goods that also proved to be financially sustainable.

Even though these insights were confirmed by the numerous critical experts' comments in the FAO online Forum on Food Security and Nutrition, where we launched a discussion on the theory and practice of PES in March 2013 (http://www.fao.org/fsnforum/forum/discussions/pes), many of these trends have generally gone unnoticed in the academic literature on the theory and practice of PES.

My goal then was to draw attention to this gap by selecting PES cases in which public–private partnerships enabled local entrepreneurship and innovation and thus ensured the financial sustainability of the provision of environmental services in countries where there was no government funding to compensate farmers for the positive externalities of sustainable agriculture.

All these case studies were presented in a multistakeholder dialogue at FAO on September 12 and 13, 2013 (http://www.fao.org/nr/aboutnr/incentives-for-ecosystem-services/case-studies/en/).

Since I assumed the position of Director of the Center for Corporate Responsibility at the University of Zurich in October 2013, I was unable to continue my work as project coordinator.

My successor did not further pursue the idea that entrepreneurship and innovation can be a source of positive externalities in PES schemes. This gave me the opportunity to continue with the work on a personal basis. I then finally decided in summer 2014 to write a book mostly based on the draft working papers that had already been favourably reviewed by members of the steering committee of the project at FAO. I am therefore very grateful to my colleagues at FAO who provided me with valuable feedback on the working papers, especially Tomas Lindeman, Rao Matta, Julien Custot, Alexander Mueller, Daniela Ottaviani, Leslie Lipper, Cassandra DeYoung, Alberto Sandoval, Pierre Gerber, Bernadette Neves and Stéphane Jost. I am also very thankful to all the PES experts I met in Kenya (see also Annex).

In Switzerland, I would like thank Hans-Jörg Lehmann from the Swiss Federal Office of Agriculture and Michel Dumondel from ETH Zurich for supporting me during my time as project leader at FAO, and Susan Kaplan and Isabelle Schluep for her valuable assistance with the preparation of the final manuscript.

Finally, my thanks go to Johannes Glaeser from Springer Publishing and the reviewers of the Series on CSR, Sustainability, Ethics and Governance.

Philipp Aerni
Center of Corporate Responsibility and Sustainability
University of Zurich

Testimonials

This book is a refreshing challenge to conventional approaches to payments for environmental services. It offers a more realistic framework that accounts for human creativity and innovation in environmental management.
Professor Calestous Juma, Harvard Kennedy School, Cambridge, MA, USA (Author of the "The New Harvest: Agricultural Innovation in Africa", Oxford University Press 2011)

Philipp Aerni offers a most useful critical assessment of the theory and practice of payments for ecosystem services. His book helps us understand why such payments sometimes fail to reach their objectives, and what could be done about it, particularly in less developed countries.
Professor Thomas Bernauer, ETH Zurich, Switzerland

Drawing on lessons learned from case studies in Kenya, this important book shows what is needed to improve environmental services and sustainability in the developing world.
Paul Slovic, President, Decision Research and Professor, University of Oregon, Eugene, OR, USA

This book vividly assesses the use and impact of PES (Payments of Environmental Services) in agriculture. Addressing the gap between PES theory and practice in the historical context, valuable lessons are learned from case studies in Kenya. They assist in refining future policies, linking PES to new business opportunities. The book is essential reading in the quest for squaring increasing food production and sustainable development and for the design of future rules of the international trading system.
Thomas Cottier, Professor of Law, World Trade Institute, University of Bern

Executive Summary

In 2012, the Rio+20 conference concluded that sustainable land use systems play a crucial role in enabling a global green economy. These systems must cope with the double challenge of increasing the production of primary commodities for a growing and more affluent world population while halting the continuous loss of vital ecosystem services as a result of unsustainable practices. The growing global demand for fuel, fibre, feed and food must therefore be met through sustainable intensification techniques that enable production to be increased without further impairing the provision of valuable environmental services.

This book makes the case that current environmental policy instruments, such as Payments for Environmental Services (PES), may not be able to cope adequately with this doubly challenge because they are primarily focused on regulating unsustainable change in developed countries but may actually hamper efforts to facilitate sustainable change in developing countries. These policy instruments, derived from neoclassical welfare economics, are based on the implicit assumption that there is a trade-off between economic development through technological change (increasing provisioning services of ecosystems) and the conservation of natural resources (protecting the supporting and regulating services of ecosystems). This framing of the global environmental challenge shaped the research focus as well as the conclusions of the Millennium Ecosystem Assessment published in 2005 (MEA 2005). The final report of the MEA advocated PES as an effective way to internalize the environmental costs of economic activities.

Emerging economies that must cope with the many social and environmental challenges of economic growth are undoubtedly challenged by trade-offs between economic development and the conservation of natural resources. But they invest as much in technological change as they experiment with different types of PES approaches to cope with the challenges. In developed countries, PES schemes have been built into national agricultural and environmental policies, but the generous payments have often undermined incentives to actively invest in innovation that improves the quality of environmental services while being economically viable. Building PES schemes into national policies has proved to be challenging in least developed countries where governments cannot afford to dispense generous payments for the provision of environmental services and where poverty—not affluence—turns out to be the main enemy of the environment.

We therefore need a more comprehensive and, at the same time, differentiated theory on the effective provision of environmental services that is based on concrete experience on the ground. This would lead to context-specific approaches that ensure the sustainable use of valuable ecosystem services while incentivizing innovation-driven local entrepreneurs to invest in new markets for environmental goods and services that generate positive welfare effects for society and the environment. Such approaches require a profound understanding of the local context. Local stakeholders must therefore be in charge of implementation and they must be able to draw on government support. Governments must actively help mobilize the relevant competences and resources available in business, civil society and academia to ensure that local actors are successful in their efforts to improve the provision of environmental services while also creating new economic opportunities for those who depend on the use of natural resources—in least developed countries these are often poor semi-subsistence farmers. In other words, governments need to become facilitators of change and not merely regulators of unsustainable change. They must support local stakeholders in their efforts to respond to environmental regulation with innovation.

The Current Problem with PES

Projects that involve PES in one way or another are very popular with national and international organizations involved in environmental policy and management and they are the subject of countless academic publications that evaluate many different types of incentives and the effectiveness of PES applied to many different ecosystems services and many different socioeconomic environments. But they mostly tend to stick to the same neoclassical baseline assumption that pure market-based institutions are able to assign a measurable value to environmental services and thus create a market in which this value is properly taken into account. The sustainable provision of environmental services in today's global knowledge economy, however, should not be based on externally assigned buyers and sellers of environmental services and depend on large amounts of subsidies that keep such fictitious markets alive. This explains the growing trend towards hybrid PES projects that are no longer guided by classic principles of PES theory.

The Purpose and Content of the Book

This book discusses the gap between PES theory and PES practice by means of a literature review and illustrates it by means of case studies based on field visits to PES projects in Kenya in August 2012. Assessment of PES projects in developed and developing countries indicates that once the projects are implemented, they look completely different from what would have been predicted by PES theory.

They often fail to be financially sustainable because of lack of incentives for the local people to continue with activities that were introduced by an external actor. But sometimes these projects create unexpected opportunities for local entrepreneurs to create a new market for environmental goods through innovation. These opportunities arise from indirect effects of PES projects, such as enhancing the social and business network of local people, as well as improving their access to new investment, capacities and technologies. In this way they become innovative green entrepreneurs who have realized that new regulation, rather than just restricting existing business activities, can also create opportunities for new ones, if appropriate facilitating policies are in place to encourage investment in green innovation and green markets.

This book consists of three parts that address the challenges of market-based instruments (MBIs) in environmental policies and projects from a historical, a theoretical and a practical and environmental perspective (based on field visits to projects in Kenya). In this context, the focus will be on the emergence of the concept of PES, how it has evolved over time and space and how the political framing also affects research priorities as well as the theory and practice of PES.

(a) The historical perspective

The historical perspective illustrates the evolution of PES schemes through documented cases from developed and developing countries over the past four decades. It shows that PES has its roots in the first environmental regulatory framework established in the United States (USA) in the 1970s. At that time, the USA designed laws to protect biodiversity, water and soil fertility. They required landowners either to offset the negative environmental externalities generated through a particular land use or to leave land fallow to avoid negative externalities from agriculture in the first place. In Europe, PES eventually became an important tool not just in environmental policy but also in efforts to redirect agricultural support from production-tied subsidies to direct payments aimed at promoting multifunctional agriculture after the end of the Cold War. 'Multifunctionality' was meant to address the non-trade concerns in agriculture that would result from a more open and transparent agricultural trading system as envisioned by the World Trade Organization (WTO)'s Agreement on Agriculture (AoA). The concept of multifunctional agriculture argues that farmers not only produce food but also provide other important services for society and the environment that are not reflected in the price of the primary product they produce. Since consumers and taxpayers largely supported this view, the new system of direct payments (direct income support for farmers, not meant to be linked to production decisions) was framed as a form of remuneration for farmers who adopt sustainable agricultural practices and thus contribute to the provision of environmental services. The PES scheme in Europe is therefore linked to the creation of positive externalities, rather than the offsetting or avoidance of negative externalities in agriculture, which is the focus of US agri-environmental policies.

Eventually, PES also took root in many developing countries as governments started to implement more stringent environmental regulation and to provide incentives for landowners to adopt conservation practices in the 1990s. The case of Costa Rica illustrates how these early PES schemes were mainly designed to restrict the use of natural resources, such as forests, which are providers of ecosystem services that have considerable economic value. Offering support for PES projects and policies in the developing world became increasingly popular with donors in developed countries who often attempted to replicate what they believed to work in their home country. Yet PES schemes in developed countries tended to be designed to make agriculture more extensive and to ensure decentralized settlement. This makes sense for a capital-intensive agricultural sector that has already reached an advanced stage in the process of structural change. But it may not work in least developed countries (LDCs) where structural change has yet to take place. In other words, LDCs need economic and technological change in agriculture if they are to become sustainable, whereas the agricultural policies in developed countries are largely designed to protect farmers from such a change. False baseline assumptions have their roots in the theory of PES and are not just a problem in the concrete implementation of PES projects.

(b) The theoretical perspective

The second part of the book deals with the theoretical foundations of PES that are derived from neoclassical welfare economics. This economic theory also has its roots in the Cold War period and largely shaped agricultural, environmental and development policies in Europe and the USA in the 1970s. A major weakness of neoclassical welfare economics is that its comparative-static equilibrium models are unable to take into account the welfare effects that are generated when the private sector makes use of new knowledge to produce new goods and services that lead to the more sustainable use of natural resources. As a consequence, the value of public–private partnerships in PES schemes as well as the role of innovation and entrepreneurship for the sustainable management of natural resources tends to be neglected. The failure to take these dynamic aspects of PES into account is particularly problematic in the global knowledge economy of the twenty-first century, which provides many new opportunities to generate increasing revenues and employment through the development and introduction of green goods and services. After all, knowledge is the only resource that is not scarce and the failure to take advantage of new knowledge to develop new solutions, technologies and products also has a negative long-term impact on global sustainability.

The comparative-static view that prevails in PES theory may explain why the business case for PES is often based on shaky foundations. The assumption that there is a market consisting of a clearly identifiable buyer (beneficiary) and a seller (provider) of environmental services hardly ever applies in reality where PES projects are initiated and funded largely by private intermediaries or the state. PES becomes even more problematic when it is aimed to address poverty and environmental problems at the same time. As will be illustrated in this book, this often leads to considerable distortions, converting the polluter-pays principle (the principle that is supposed to guide progressive environmental policies) into a sort of pay-the-polluter principle where unsustainable polluting farms become eligible for payment in return for adopting more sustainable practices while farmers who already run sustainable farms tend to be ignored. This is particularly true of LDCs where it is not economic development but poverty (coupled with population growth and limited economic prospects) that is the main enemy of environmental sustainability.

Many new PES schemes have aimed to address these shortcomings with mixed results. Concrete practical experience indicates that PES projects that proved to be effective and financially sustainable have achieved this for reasons that were not anticipated by PES theory. These are the findings of the field studies of PES- and PES-related projects in Kenya.

(c) The practical/environmental perspective

The field visits to Kenya took place from August 27 to September 7, 2012. The projects studied show how efforts to simultaneously improve farm productivity, rural livelihood and environmental services tend to be linked to hybrid forms of PES. Such hybrid forms include in-kind payments, price premiums for sustainable production, special input discounts for complying with sustainable practices and business opportunities (e.g. sale of tree seedlings or conversion of waste products into farm inputs and energy products). These locally organized markets, which often emerge in response to new social networks created by a PES scheme, as well as new environmental regulation and standards, are more important for the local participants than the actual transfer payment for the provision of a particular environmental service. Business opportunities linked to PES schemes also tend to improve the environmental performance of such schemes since they encourage continuous efforts to improve measures and techniques to enhance environmental quality in the expectation of being rewarded through more investment, special discount offers on inputs and better access to output markets. A challenge in all types of PES- and PES-related projects are the high transaction costs of monitoring, reporting and verification (MRV) and the lack of instruments that can exactly measure the improvements in water quality, biodiversity or carbon sequestration resulting from the adoption of sustainable agricultural practices. Projects that succeeded in substantially lowering the costs of MRV—and in measuring the provision of the environmental services more accurately—did so because they made effective use of new technologies, especially ICT. The Kenyan field studies also indicate a trend away from classical PES schemes towards broader efforts to remunerate or reward farmers who generate positive externalities in agriculture.

The concluding chapter of the book combines the insights gained from the three perspectives to make policy recommendations on how to improve environmental services not only through regulation but also through incentives to innovate. The role of governments as facilitators of sustainable change will be crucial in efforts to build a global green economy that embraces and improves the linkages between rural and urban development.

Abbreviations

AECF	Africa Enterprise Challenge Fund
AFOLU	Agriculture, Forestry and Other Land Use
AGRA	IFAD Alliance for a Green Revolution in Africa
AoA	WTO Agreement on Agriculture
ASDS	Agricultural Sector Development Strategy
AWG/LCA	Ad Hoc Working Group on Long-term Cooperative Action under the Convention
B2B	Business to Business
B2C	Business to Consumer
BSCI	Business Social Compliance Initiative
CAP	EU Common Agricultural Policy
CARE	Cooperative for Assistance and Relief Everywhere
CBD	UN Convention on Biological Diversity
CCT	Conditional Cash Transfer
CDE	Centre for Development and Environment, University of Bern
CDM	Clean Development Mechanism
CFA	Community Forest Association
CIFOR	Center for International Forestry Research
COP	Conference of the Parties
CoP	Code of Practice
CPCC	Control Points and Compliance Criteria
CSA	Climate-Smart Agriculture
CSR	Corporate Social Responsibility
CWA	US Clean Water Act
EPA	US Environmental Protection Agency
ER	Emissions Reduction
ESA	US Endangered Species Act
ETI	Ethical Trade Initiative
ETS	Exchange Traded Schemes
FAO	Food and Agriculture Organization of the United Nations

FCCC	UN Framework Convention on Climate Change
FDAC	Focal Development Area Committee
FFP	Fair Flower Fair Plants
GAP	Good Agricultural Practices
GATT	General Agreement on Tariffs and Trade
GBM	Green Belt Movement
GEF	Global Environment Facility
GiZ	*Deutsche Gesellschaft für Internationale Zusammenarbeit* (German Group for International Cooperation)
GLOBALG.A.P.	Global Good Agricultural Practices
GM	Genetically Modified
GMO	Genetically Modified Organisms
GSCP	Global Social Compliance Programme
GWA	Green Water Credits
HEBI	Horticulture Ethical Business Initiative
ICC	International Code of Conduct
ICIPE	International Centre of Insect Physiology and Ecology
ICRAF	World Agroforestry Centre
IFAD	International Fund for Agricultural Development
ILO	International Labour Organization
ILRI	International Livestock Research Institute
IPBES	Intergovernmental Platform on Biodiversity and Ecosystem Services
INRA	Institute National de la Recherche Agronomique (France)
KACP	Kenyan Agricultural Carbon Project
KARI	Kenya Agricultural Research Institute
KFC	Kenya Flower Council
KWS	Kenya Wildlife Service
LANAWRUA	Lake Naivasha Water Resource Users Association
LDC	Least Developed Country
LiAPS	Livestock inclusive Agricultural Production Systems
LULUCF	Land Use, Land Use Change and Forestry
MAC	Market Access Company
MBI	Market-Based Instruments
MEA	Millennium Ecosystem Assessment
MFI	Micro-Finance Institutions
MKEPP	Mount Kenya East Pilot Project
MoA	Kenya Ministry of Agriculture
MRV	Measurement, Reporting and Verification
NAMA	Nationally Appropriate Mitigation Action
NEMA	National Environmental Management Authority, Kenya
NGO	Non-Government Organization
NNP	Nairobi National Park
NRMP	Natural Resource Management Plan

OECD	Organization for Economic Co-operation and Development
PAP	Principal Agent Problem
PES	Payments for Environmental Services
PRESA	Pro-poor Rewards for Environmental Services in Africa
PROFIT	Programme for Rural Outreach of Financial Innovations and Technologies
PWC	Payments for Wildlife Conservation
REDD	Reduction of Emissions from Deforestation and Degradation
RISEMP	Regional Integrated Silvopastoral Ecosystem Management Project
RKN	Rural Knowledge Network
RPE	Remuneration of Positive Externalities
SACCO	Savings and Credit Cooperatives
SALM	Sustainable Agricultural Land Management
SBSTA	Subsidiary Body for Scientific and Technological Advice
SLCP	Sloping Lands Conversion Program
SWC	Soil and Water Conservation
TNC	The Nature Conservancy
TPC	Third-Party Certification
TSS	Transaction Security Services
TWF	The Wildlife Foundation
UNCED	United Nations Conference on Environment and Development
UNCTAD	United Nations Conference on Trade and Development
UNEP	United Nations Environment Programme
UNIDO	United Nations Industrial Development Organization
UTaNRMP	Upper Tana Catchment Natural Resource Management Project
VCR	Verified Carbon Standard
VER	Verified Emissions Reductions
WLP	Wildlife Lease Program
WRMA	Water Resource Management Authority
WRUAs	Water Resource Users' Associations
WTO	World Trade Organization
WWF	World Wildlife Fund

Contents

List of Figures

List of Boxes

Chapter 1
The Historical Context of Payments for Environmental Services: A Trend Towards Public–Private Partnerships

Abstract This chapter discusses the history of PES in developed and developing countries and illustrates its challenges and opportunities by means of concrete cases. The cases include projects designed to provide different environmental services (clean water, carbon sequestration, biodiversity preservation, etc.) applying different public PES schemes as well as public private partnerships. The experience indicates that the lack of financial sustainability of PES projects is a major reason for the reversibility of environmental improvements once external funding stops. Long-term financial sustainability can however be achieved, if PES projects allow for the creation of markets for environmental goods driven by innovative local entrepreneurs. Such hybrid PES schemes may be especially adequate for developing countries that face financial constraints on different levels.

The linkages between agriculture and the environment are complex. Some regard agriculture primarily as a threat to environmental services while others believe that sustainable farming practices must become an integral part of the provision of these services. Those who agree with the former view tend to see payments for environmental services (PES) as a tool to compensate farmers and pastoralists for restricting their economic activities in environmentally valuable areas. Those who endorse the latter view, however, make direct payments to land users conditional upon the adoption of sustainable practices that mitigate the impact of their activities on the environment and thus help ensure the provision of environmental services. The history of PES can therefore largely be divided into agri-environmental policies that tend to be either 'use-restricting' (focused on the avoidance of negative externalities) or 'asset-building' (focused on the creation of positive externalities). The use-restricting policies have their origins in the environmental legislation passed in the United States in the 1970s, while the asset-building policies are linked to the concept of multifunctional agriculture that is aimed at remunerating farmers for the adoption of sustainable practices. This concept evolved in the 1990s in Europe and must be partially understood as a response to the geopolitical changes after the end of the Cold War and the adoption of the Agreement on Agriculture (AoA) in the General Agreement on Tariffs and Trade (GATT) in 1994. As a consequence of the end of the Cold War, policy makers found it more difficult to protect farming in the name of national security, and the interests of taxpayers and consumers had to be considered to a much greater extent in agricultural policy. As a

P. Aerni, *The Sustainable Provision of Environmental Services*, CSR, Sustainability, Ethics & Governance, DOI 10.1007/978-3-319-19345-8_1

result, farm support became increasingly decoupled from agricultural production. This move coincided with a general willingness of taxpayers and consumers to continue to support the farming sector through a system of direct payments.

In developing countries, governments started to introduce PES in environmental regulation in the 1990s as a use-restricting tool, mainly to discourage deforestation. The Millennium Ecosystem Assessment report (MEA), published in 2005, identified PES as possible tool to encourage the more sustainable use of natural resources. As a result, attention in developing countries also shifted towards PES schemes that generate positive externalities in agriculture, especially with regard to the provision of watershed services that are widely considered to have public-good character. This induced many governments in developing countries to pass legislation that supported the adoption of sustainable agricultural practices. However, most projects that remunerate farmers for the generation of positive externalities are funded by donors from affluent countries either indirectly via international funds or directly through national programmes of development and cooperation. The same applies to the development of voluntary carbon and biodiversity markets that are aimed at avoiding changes in the use of land containing valuable ecosystems (use-restricting). They allow farmers in developing countries to sell carbon or biodiversity offsets to affluent buyers in developed countries, who wish to reduce their environmental footprint.

1.1 Introduction

The current policies related to remuneration of positive externalities (RPE) and PES in different parts of the world must be viewed in the historical context of agricultural and environmental policies in which RPE/PES schemes have emerged over the past two decades (Gómez-Baggethun et al. 2010). This allows a better understanding of the wide range of RPE/PES schemes currently applied in Organisation for Economic Co-operation and Development (OECD) and non-OECD countries.

The following chapter is divided into two main parts, covering RPE/PES policies in developed countries and RPE/PES policies in developing countries. The idea of such policies, either to remunerate farmers for the creation of positive externalities (asset-building) or to compensate them for the avoidance of negative externalities (use-restricting), had its roots not only in public environmental concerns, but also in the need for developed countries to redefine the role of subsidies in agriculture after the completion of the Uruguay Round, which ended in the GATT in 1994, and eventually the World Trade Organization (WTO). This Round was able to reach a consensus among the GATT member states on reductions in agricultural support and tariff protection with certain exemptions and safeguards. One of the main purposes of the resulting Agreement on Agriculture (AoA) of the WTO was to encourage developed countries to shift their large production-tied subsidies as far as possible into the so-called ‘green box’, defined in Annex 2 of the AoA as domestic support that must not distort trade.

Since the remuneration or compensation of farmers for the provision of public-good services was not considered to be trade distorting, developed countries were able to redefine the role of subsidies as 'green box measures' (e.g., income support) and thus reassured their domestic farmers that they would not suffer from substantial income losses if they were to become subject to the new disciplines of the WTO. At the same time, domestic taxpayers and consumers could be reassured that non-trade concerns about domestic farming were being taken into account by rendering a growing share of direct payments contingent upon the adoption of predefined sustainable agricultural practices that would support the public-good character of agriculture (Aerni et al. 2011). The idea of paying farmers for the provision of environmental services must therefore also be understood as a political compromise in efforts to gain public support for a new agricultural policy after the ratification of the GATT. The theory of PES as developed by welfare economists at universities was subsequently embraced in public policy because it provided an additional rationale for the new agricultural policy.

In the section on RPE/PES in developed countries we discuss this important transition in domestic agricultural policy and the new focus on agricultural sustainability. However, in many cases, this transition from production-tied subsidies to direct payments has not led to a reduction in public expense. Further shifts from mere direct payments to eco-payments face increasing political resistance from farmer organizations and have encountered financial constraints as a result of the economic crisis in 2008. This has forced many countries to rethink their agricultural policy. We document this trend by discussing some successful cost-saving but effective public–private partnerships that introduced innovative RPE/PES schemes in Europe and the United States and the new efforts in many OECD countries to highlight the important role of innovation as a source of positive externalities in agriculture.

The same trend can also be observed in developing countries as will be documented in the second part of the chapter. Many developing countries designed regulations to minimize the impact of agriculture on the environment but there were often insufficient financial and human resources to implement these regulations. In response to the limited means of the public sector, many private initiatives (Non-Governmental Organizations (NGOs), private sector) emerged with the purpose of setting up RPE/PES schemes that would meet the goals of national environmental policies in a cost-effective way. Private sector involvement in PES schemes often turns out to be crucial in ensuring financial sustainability as will be illustrated by means of the selected case studies.

The situation in developing countries is also different from the one in developed countries because of the many international environmental agreements, such as the UN Framework Convention on Climate Change (FCCC) and the UN Convention on Biological Diversity (CBD), that all include a PES mechanism in one form or another. These agreements call upon developed countries to offer technical and financial assistance to developing countries in their efforts to comply with the new global environmental objectives. This has led to new markets, especially in the area of climate change mitigation and biodiversity conservation, where actors in the private sector invest in carbon and biodiversity offsets that can then be traded and exchanged in cap-and-trade markets or be used to obtain particular permits to build new plants or pollute/ tap water, for example, in different areas of interest.

However, since the publication of the UN Millennium Ecosystem Assessment (MEA) (2005) and its finding that agriculture could play a positive role in the provision of ecosystem services, the focus in developing countries has shifted to some extent. There has been a move away from using PES in agriculture primarily as a means to avoid negative externalities for the environment, towards making PES one of the tools that remunerate farmers for the generation of positive externalities, for example, through the adoption of sustainable agricultural practices. In this context, many watershed PES projects have been created that encourage upstream farmers to adopt sustainable agricultural practices in return for payments from the downstream water users that appreciate the ecosystem service of 'clean water'. However, as will be briefly illustrated here (and more extensively discussed in the chapter on practical perspectives derived from concrete case studies), such RPE/PES schemes in developing countries run the risk of not being financially sustainable because most of the funding to start, operate and maintain a PES project comes from the mediating foreign agent rather than the local private sector. The problem of financial sustainability, however, can be addressed effectively by making use of PES as a vehicle to create innovative local markets for environmental goods, as will be illustrated later.

The chapter concludes that there is a trend towards encouraging innovation and public–private partnerships in RPE/PES schemes in developed and developing countries that, overall, may contribute to a higher level of financial sustainability of such schemes. Several conditions need to be met, however, to ensure that these schemes reach and empower the poor.

1.2 PES in Developed Countries

PES schemes in OECD countries have grown out of existing agricultural subsidies schemes and national environmental regulation that were mostly put in place during the Cold War period to ensure national food security on the one hand and sustainable natural resource management on the other. With the fall of the Berlin Wall in 1989 and the successful completion of the Uruguay Round of the GATT in 1994, it became harder to politically justify the highly subsidized and protected agricultural sectors to ensure food security. In turn, concerns about the environment, food safety and the future of family farming became increasingly important to affluent taxpayers and consumers (Aerni 2009). This eventually caused a shift in agricultural policy, especially in Europe, from production-based subsidies to subsidies designed to address non-trade concerns such as food safety, rural development and the environment.

However, this did not mean that there was a consensus on how to achieve these valued public goods cost-effectively and on how to define the desired quantity of such goods. For example, the United States and the European Union (EU) adopted different approaches to address the environmental externalities created by agricultural production. In both regions, agri-environmental programmes have been designed to transfer income to farmers while mitigating the impact of modern agriculture on the environment. Agri-environmental policies in the two regions, however, build on different baseline assumptions about the fundamental relationship between agriculture and the environment. The agri-environmental programmes in the USA reflect the view that there is an intrinsic conflict between the goals of expanding (or maintaining) agricultural production and preserving the environment. The policies in Europe, by contrast, assume that the expansion of agricultural activity can actually benefit the environment, provided that expansion is undertaken in an appropriate and sustainable manner (Baylis et al. 2005).

1.2.1 Agri-Environmental Policies in the United States

The 1973 Agriculture and Consumer Protection Act represented a major shift in US agri-environmental policies from mere agricultural supply management towards the more comprehensive management of farming and its environmental and social challenges. It adopted target prices and deficiency payments as a tool to support farm income, authorized disaster payments and disaster reserve inventories, and created the first rural environmental conservation programme, among other things. Since then, Farm Bills[1] have put ever more emphasis on the conservation of environmental services. The 2008 Farm Bill supported multiple, mostly voluntary, schemes that were dedicated to the conservation and sustainable management of ecosystem services.[2]

In addition to the Agriculture and Consumer Protection Act, two crucial environmental acts, passed in the 1970s also had a profound impact on policies designed to protect wetlands and biodiversity: the Clean Water Act (CWA) and the Endangered Species Act (ESA). These acts eventually led to the creation of markets for biodiversity, initiating a trade in wetland and biodiversity offsets.

Since the mid-1980s, a series of biodiversity markets worth more than US$3 billion a year have been created in the United States (Worldwatch Institute 2008). Although these are markets involving the private sector, it is ultimately the

[1] The farm bill is an omnibus, multi-year piece of authorizing legislation that governs an array of agricultural and food programs.

[2] These schemes include (e.g. Agricultural Management Assistance Program, Chesapeake Bay Watershed Initiative, the Cooperative Conservation Partnership Initiative, the Conservation of Private Grazing Land Program, the Conservation Reserve Program, the Conservation Stewardship Program, the Environmental Quality Incentives Program, the Agricultural Water Enhancement Program, Conservation Innovation Grants, Farm and Ranch Lands Protection Program, Grassland Reserve Program, Healthy Forest Reserve Program, Small Watershed Rehabilitation Program, Wetlands Reserve Program, Wildlife Habitat Incentive Program).

government that makes the markets possible through environmental regulation. On the one hand, such markets have led to investments in environmental goods and services that help mitigate the impact of economic development on the environment; on the other hand, they are also subject to criticism from developers as well as from conservationists, as will be illustrated with the following two examples [see Business and Biodiversity Offsets Programme (BBOP) (2013) for further information].

1.2.2 The Delhi Sands Flower-Loving Fly: What Is the Exact Value of An Endangered Species?

An example of how a market for biodiversity might evolve in response to the Endangered Species Act (ESA) is Colton, a small town nestled in the sand dunes in San Bernardino County in California. It is one of the fastest-growing counties in the United States—but, at the same time, provides a habitat for an important endangered species, namely, the Delhi Sands flower-loving fly (*Rhaphiomidas terminatus* abdominalis), a mydid fly in the genus Rhaphiomidas that was emergency-listed by the US Fish and Wildlife Service on September 23, 1993. It became the first fly—and only the seventeenth insect—to be declared an endangered species in the United States. According to the US ESA, no individual or entity, public or private, can harm an endangered species without a permit from the government. As a result, the construction of a hospital in San Bernardino County came to a halt because it would have paved over seven acres of occupied fly habitat. This was all of a sudden illegal. What makes this flower-loving fly so valuable? It is a pollinator that has important and symbiotic relationships with plants. However, the accurate assessment of its exact ecological value is not possible owing to the complexity of dynamic and highly interconnected ecosystems. Having determined that the fly should be protected, the State of California decided to let the market decide what it costs to conserve it. The market rate was assessed to be between US$100,000 and US$150,000 per acre. This meant that costs amount to a maximum of US$150,000 to create a habitat for the Delhi Sands flower-loving fly. This amount could then be used to protect or create a habitat for that same fly somewhere else (considered to be an 'offset'). The main aim of such offsets is to avoid a net loss of biodiversity associated with development, particularly urban and coastal development. Yet, the concept of offset not only faces criticism from the private sector for the rather arbitrary price-setting of the value of biodiversity, but it is also widely criticized by environmentalists who reject the narrow definition of biodiversity, its inability to capture the complexity and diversity of ecosystems, and the lack of resources for implementation and long-term monitoring (Burgin 2008). This may not apply only to biodiversity, but also to wetland mitigation.

1.2.3 Wetland Mitigation Banking as a Consequence of CWA

There are private, for-profit, wetland mitigation bankers in the United States who make money by creating, enhancing, and restoring wetlands and then selling the resulting 'wetland credits' to developers in search of offsets. The bankers focus on buying wetland areas that are likely to experience economic growth; subsequently, they obtain 'credits' for the 'creation, enhancement, and restoration' of wetlands (hence creating a 'wetland bank') which they then sell to developers who find themselves in need of compensation. How could such a market have evolved? It is largely a response to the CWA. Section 404 of the CWA is designed to prevent the placement of dredged and filling materials into the 'waters of the US'. As a consequence, anyone wishing to dredge or fill a wetland considered as being of national importance in the United States must first obtain a permit through a programme administered by the US Army Corps of Engineers and the US Environmental Protection Agency (EPA). In deciding whether to award this permit, the EPA and the Corps first determine whether the damage to the wetlands can be avoided. If it cannot, the next step is to minimize the damage. Finally, the developer is supposed to offset, mitigate, or compensate for any damage that cannot be minimized. The law is also quite clear on what is considered appropriate compensation for the damage to wetlands: developers must 'create, enhance, or restore' an amount equal to or greater than the amount being damaged in a wetland of 'similar function and value' in the same watershed. In some special cases, protecting a similar wetland is considered to be a suitable compensation, though this is rare. The compensation for any development projects that harm wetlands—whether done by private developers or the government—can be undertaken by the developers themselves or by third parties. The Army Corps of Engineers and the EPA are charged with overseeing this process and making sure the compensation happens.

In other words, wetland mitigation banking is possible because the government is restricting supply—allowing the market to set a price (a value) on this particular aspect of biodiversity. Wetland banking is estimated to cover 24,178 acres and to generate around $1.1–$1.8 billion in revenues.[3] Entrepreneurial wetland mitigation bankers account for about one third of that business. In order to ensure the quality and longevity of wetlands, the Army Corps of Engineers and the EPA require wetland bankers to provide both legal and financial assurances that the 'created, enhanced, or restored' wetland will last (presumably) in perpetuity. The legal assurances are usually provided through conservation easements (legal restrictions on the use of land) held by third parties (usually a non-profit or the government). The financial assurances can take a variety of forms. They are either trust funds set up to produce the interest necessary to run the bank or bonds or letters of credit that hold the bank financially liable for the protection of the wetlands. Wetland mitigation banking requires a considerable amount of enforcement and verification. The government

[3] See http://www.ecosystemmarketplace.com/pages/dynamic/web.page.php?section=biodiversity_market&page_name=uswet_market

agencies overseeing the system have to monitor continuously and ensure that the promised wetland protection is delivered. Such ‘perpetual oversight’, however, is costly and usually very difficult for understaffed and underfunded government agencies to provide (Bayon 2008).

1.2.4 Agri-Environmental Policies in Europe: Remuneration of Positive Externalities

The Common Agricultural Policy (CAP) of the EU started as a farm support scheme in 1962 and largely reflected France’s preference for state intervention in agriculture. The policy was guided by three major principles: market unity, community preference and financial solidarity—all directly or indirectly designed to ensure farm support for increasing agricultural production in Europe. Efforts to reform the CAP are almost as old as the CAP itself but resistance to reform by the beneficiaries of the system prevailed until pressure increased when agriculture also became subject to GATT disciplines with the successful completion of the Uruguay Round in 1994. The policy shift towards multifunctional agriculture took concrete shape with the CAP Agenda 2000 when several rural development measures and agro-environmental schemes were introduced in member states. In 2003, the fundamental reform of the CAP finally took place when subsidies were ‘decoupled’ from particular crops. Direct farm income support (the Single Farm Payment) was however subject to ‘cross-compliance’ conditions relating to environmental, food safety and animal welfare standards. The multifunctional character of agriculture, taking into account also its contribution to the sustainable management of ecosystem services, thus became officially enshrined in EU agricultural policy. The concept of multifunctionality recognizes the positive externalities that agriculture generates for society and the environment (e.g., protection of the environment, preservation of landscapes and agrobiodiversity, safe food, socioeconomic and cultural well-being of farm families, recreation value for urban residents). It is assumed that all these public goods may not be valued in the global market for agricultural commodities but reflect the concerns, preferences and values of taxpayers and consumers in affluent countries (Renting et al. 2009). Yet, despite the success of CAP reforms in reducing agricultural export subsidies and the overall budget share of agriculture in the annual EU budget (from 71 % in 1984 to 39 % in 2013), it is not clear to what extent the reforms are really contributing to the improvement of environmental services (Kleijn et al. 2006). Moreover, policy incentives to move towards more extensive agricultural practices have resulted in lower annual agricultural productivity growth rates in the past decade. This may become a problem in view of the rising demand for food, feed, fibre and fuel in the emerging economies (Noleppa et al. 2013). Since 2013, the CAP has therefore been seeking to reconcile the need to ensure global food security with the challenge of ensuring the sustainable management of environmental services. It aims to do so by giving more emphasis to private initiatives and innovation in rural areas.[4] A trend towards public–private partnerships in the sustainable management of ecosystem services has been observed in Europe as well as the United States.

1.2.5 Public–Private Partnerships

As well as publicly funded schemes to support sustainable agricultural practices, interesting public–private partnerships have been set up on both sides of the Atlantic. In the United States, the *New York City Watershed* (NYC Watershed) memorandum of agreement in 1997 has been quoted as a positive example of a successful public–private partnership that ensures a safe water supply for the metropolitan region of New York (FAO 2008). The key components of the agreement were the acquisition of land, and stewardship programmes near reservoirs, wetlands and watercourses and the development and upgrading of sewage systems, water treatment plants and storm-water management facilities and stream water corridor protection. Moreover, an economic development bank was created to promote upstream development consistent with catchment protection and to support the implementation of best management practices on farm and forest land. In this context, a voluntary agreement was reached with farmers to adopt conservation practices in return for PES. The agreement was conditional upon the participation of 8 % of farms within 5 years and on achieving the goals for protection of the entire landscape rather than of individual farms. These conditions were met early on and are being successfully monitored and enforced to date by the Watershed Agricultural Council which also supports and promotes small farm businesses provided that the farm products have been produced using best management practices. The NYC Watershed protection initiative demonstrates that both downstream water quality goals and upstream economic objectives can be achieved through voluntary partnerships between the different users and implementation of community-based watershed protection. The initiative also showed that by protecting reservoirs and areas surrounding source waters it is possible to supply water for a massive urban population without the need for expensive filtration or chemical treatment.[5]

The NYC Watershed Agricultural Program is a good example of a public-private partnership. It was developed jointly by the relevant authorities of New York City and the farm community to avoid the need for regulatory controls on agricultural operations in the watershed. The Council uses land conservation techniques such as whole farms plans, forest management plans and conservation easements to help

---

[4] For more information see: http://ec.europa.eu/agriculture/eip/documents/eip-opportunities_en.htm

[5] For more information see: http://www.asce.org/Sustainability/Sustainability-Case-Studies/New-York-City-Watershed-Management/

farmers, foresters and private landholders to address water pollution concerns on properties located in the Croton and Catskill/Delaware watersheds.

In Europe, the most outstanding example of a successful public–private partnership is probably the Vittel (Nestlé Waters) case of PES in north-eastern France. Vittel wanted farmers in the protection perimeter to transform their land into grassland to ensure that their premium label for 'natural mineral water' was not endangered. Vittel waters are characterized by a total absence of nitrites and a particularly low level of nitrates. To be labelled 'Vittel', the water cannot contain more than 4.5 mg of nitrates per litre and must not contain pesticides. However, inducing farmers to revert to an extensive farming system was simply not possible unless the right system of incentives was developed. This implied addressing the land, labour, and capital shortages all at once. Adopting a new farming system implied that the farmers were willing to change under a number of mutually agreed conditions. Farmers' unions and other agricultural organizations were concerned that the changes would disturb the local farm economy and some farmers influenced by them were reluctant to accept the contract. On the other hand, because Nestlé Waters Vosges (Vittel, Contrex and Hépar) is a major employer in the basin (1800 jobs for a population of around 10,000 in Vittel and Contrexéville) and water quality is a key factor, the proposal appeared legitimate to farmers who could see the benefits of maintaining the Vittel business. Many farmers had family members who were employed by Vittel, and local development was at stake in a region where unemployment had been rampant for decades. A major step was taken in 1992 when Nestlé Waters, then full owner of Vittel, created Agrivair, an intermediary responsible for negotiating and implementing the programme. Agrivair was strategically located just outside the town of Vittel, close to farmers and farmers' associations.

Nestlé Waters' ability to inject large amounts of money at the beginning of the programme was fundamental to getting it started. But the programme succeeded for reasons beyond financial ones. The programme would not have been possible without listening to farmers, establishing a permanent dialogue with them, and understanding their perspectives—not only in terms of farming practices but also in terms of life choices. The entire programme was essentially a learning-by-doing experiment that involved not just numerous sellers (upstream farmers), a large buyer and investor (Vittel/Nestlé) and a trusted and effective mediator (Agrivair) but also a multidisciplinary research team from the Institute National de la Recherche Agronomique (INRA). To fine tune recommendations made to farmers, INRA monitors the nitrate concentrations all year round at 17 sites across four soil types and two types of farming systems. Agrivair monitors the farming practices, the sustainable use of new building facilities, and the livestock stocking rate. Farmers' compliance with the new extensive farming system proved to have an almost irreversible effect since once they had made the switch, farmers had no incentive to revert to former practices. It illustrates how a 10-year process was necessary to transform conflict into a successful partnership (Perrot-Maître 2008).

The Vittel programme illustrates what a direct PES scheme would look like in practice. Ideally, a perfect PES would be able to establish a precise link between farmers' practices and nitrate and pesticides concentrations in the aquifer—although, given the complexity of hydrogeological relationships, it is doubtful whether any programme could ever be expected to do this.

Agrivair has become a regional economic player itself by offering different types of agricultural services to farmers and creating other businesses in the field. It took a further step in 2012 to embed its activities into a wider rural development context and to generate value from biodiversity conservation. Its benchmarking system on biodiversity is an innovation in the domain of business-to-business labelling, which is also supported by the EU.[6] Agrivair also contributed to the creation of the Terre-Eau association, which is open to all stakeholders in the region, both public and private. Through a pooling of means (a think tank, support for project management and innovative start-ups), Terre-Eau aims to reinforce the environmental value of the region via a new co-development that is mindful of the necessity to protect water resources. Terre-Eau will also continue the actions initiated by Agrivair, such as the project for a biomethaniser that will be coupled with a protein production unit for animal feed (Agrivair 2012). All these activities will generate additional revenues and employment in the region and thus contribute to the eventual financial sustainability of the PES project.[7]

Another example of how the private sector could contribute to more sustainable practices in agriculture and thus enhance the quality of ecosystem services, such as biodiversity, is the example of Syngenta's Operation Pollinator project. The company makes most of its returns worldwide through the sale of seed and plant protection products and services. In the 1990s, however, it realized that there was a need to dedicate more research activities to sustainable agro-ecosystems, especially with regard to the shift of agricultural policies from production-based subsidies to the RPE in agriculture. On the one hand, the agro-chemical company wanted to prove that its business is compatible with the goals of agricultural sustainability; on the other hand it also saw a business opportunity in assisting farmers in their efforts to fulfil agri-environmental cross-compliance criteria on the national and the EU level to access additional direct payments conditional upon the provision of environmental services. It eventually developed the so-called Operation Pollinator project that aims to provide the essential habitat for pollinating insects and birds on farmland. Since the necessary agro-ecological conditions differ from one region to another, Syngenta collaborates with local universities and farmer groups as well as NGOs, government bodies and food chain partners to design a tailored seed package of annual wild flowers and oilseed species that attract high-value insects and birds at the fringes and corners of fields providing forage and nesting opportunities throughout the year. The additional biodiversity also helps to make crop cultivation more productive, minimize the impact of intensive agriculture on the environment and, at the same time, makes the land more attractive for agro-tourism. The biodiversity seed package, which is given to farmers either in return for a small

[6] See also http://ec.europa.eu/environment/biodiversity/business/index_en.html

[7] For the most recent information on the Vittel case see http://www.fao.org/fileadmin/user_upload/pes-project/docs/FAO_RPE-PES_Vittel-France.pdf

share of the additional payment or as part of the overall service delivery package (depending on the country), proves that the initial payment for the RPE can result in a new market for environmental goods.[8] This market allows for a certain specialization (not all farmers need to do everything themselves in order to ensure agricultural sustainability). Such markets generate local revenues as well as local jobs and thus ensure the financial sustainability of RPE/PES.[9] The lessons learned from the past 5 years with Operation Pollinator is that farmers appreciate professional assistance in their efforts to meet the conditions required to access eco-payments. The experience also shows that the private sector can contribute to the generation of positive externalities through innovation in management and product development. The great challenge will be to stimulate more competition in this new emerging market for environmental goods as well as enhancing access to these goods and services, not only in developed but also in developing countries.

1.2.6 Taking into Account the Role of Innovation in Sustainable Agriculture

The cases of Vittel and Operation Pollinator illustrate how business can contribute to sustainable local development. The private sector has much to contribute, not only in terms of financing PES, but also in terms of know-how, capacity, financial and business administration, networking, and the establishment of public–private partnerships.

Partly in response to these positive experiences with public–private partnerships, numerous OECD countries are now in the process of reforming their agri-environmental policies in the direction of more private sector involvement, decentralized forms of governance and incentives to develop and adopt innovation and innovative practices that benefit not just local society and the environment but also create new business opportunities (OECD 2011). These policies are designed to better effectively address the trade-offs between development and environmental objectives through the creation and adoption of innovation, endogenous development and incentives to meet local sustainability standards and objectives. In other words, the concept of multifunctional agriculture is being revised in the sense that economic change is no longer seen as a threat to the environment and social cohesion in rural areas, but as an opportunity (EC 2011; CH Bundesnetzwerk laendlicher Raum 2012). In this context, innovation plays a crucial role in reconciling the economic, social and environmental objectives of sustainability. It has therefore been placed at the heart of the Europe 2020 strategy for growth and jobs. With more than 30 action points, the 'Innovation Union' aims to improve

[8] For a detailed account of the Operation Pollinator project see http://www.fao.org/fileadmin/user_upload/pes-project/docs/FAO_RPE-PES_Syngenta-Europe.pdf

[9] See also http://www.cbd.int/CBDLive/media/VideoGallery/newsletter_en/oct_2010.pdf

conditions and access to finance for research and innovation in Europe, and to ensure that innovative ideas can be turned into products and services that create growth and jobs, including in rural areas.[10]

This is also one of the basic lessons learned from the New Zealand experience, where the agricultural sector was liberalized in the early 1980s in conjunction with new progressive agricultural and environmental policies that were designed to facilitate sustainable and inclusive change rather than merely regulating potentially unsustainable change (Aerni 2009). In other words, the government assumed the role of a facilitator for regional development and became a coach of entrepreneurial farmers by supporting a culture of collaboration between research institutes, local governance institutions and business that was designed to respond to the needs of farmers to cope effectively with the regional sustainability and business challenges (Aerni et al. 2009). The lesson learned from New Zealand's paradigm shift is that the objectives of multifunctional agriculture can be met effectively by abandoning the implicit baseline assumption that economic competitiveness comes at the expense of environmental and social sustainability. Instead, the experience in New Zealand shows that the new knowledge economy provides ideal conditions for farmers to empower themselves and address local environmental challenges through quicker and easier access to relevant knowledge and technology. It encourages them to use their natural resources more sustainably, to take advantage of the opportunities of technological and economic change and assume a more assertive position in negotiations with the powerful players in the global food chain (Aerni et al. 2009).

1.3 PES in Developing Countries

Governments in non-OECD countries pursued a different agricultural strategy. In view of the link between urban poverty, food insecurity, and political instability, agricultural policies tended to subsidize food prices to prevent riots by urban consumers rather than to support agricultural producers in rural areas. Farm subsidies and environmental management in agriculture did not matter much. In response to the UNCED conference in Rio in 1992, however, many governments in developing countries strengthened the role of agro-environmental institutions with the support of numerous international organizations and donors. Many stringent environmental laws have been passed to minimize the impact of agriculture on the natural environment. The need to implement such laws cost-effectively also made PES schemes attractive to less developed countries, provided that foreign aid supported such schemes in financial and technical terms (Greenspan Bell and Russell 2002). Depending on whether such foreign-funded PES schemes were designed to remunerate the creation of positive externalities within agriculture

[10] See http://ec.europa.eu/research/innovation-union/index_en.cfm?pg=home

(food security, agro-biodiversity, watershed services, cultural traditions) or the reduction of negative externalities affecting the environment outside agriculture (wildlife conservation, protection of forests, biodiversity) policies could look more similar to the use-restricting policy responses to non-trade concerns of the United States or to the asset-building policy response of Europe.

The first PES-related policy initiatives in developing countries consisted primarily of forest conservation initiatives in Latin America (FAO 2007). In this context, Costa Rica pioneered the introduction of PES-specific legislation in the 1990s (Pagiola 2008). Its Forestry Law 7575, approved in 1997, was focused on environmental protection outside agriculture (avoiding negative externalities). The purpose was to enhance various forest environmental services (e.g., clean and abundant water, carbon sequestration, biodiversity conservation, provision of scenic beauty) through compensation payments to land and forest owners (FAO 2007). Even though it is questionable whether the Forestry Law created the expected additionality—some attribute the increase in afforestation to the decline of the livestock industry in Costa Rica (Pfaff et al. 2008)—the pioneering act in Costa Rica influenced many other developing countries to pass laws that discourage deforestation through incentive-based policy schemes.

Forest conservation initiatives in developing countries represent a tool to compensate local farmers and pastoralists for not doing certain things that could negatively affect the environment, such as cutting down forest (Climate Focus 2012). They are designed to discourage agricultural activities in areas that are protected for the value of their wildlife (ILRI 2012) and for their biodiversity (BBOP 2013). In other words, many of the early initiatives in developing countries embraced a US understanding of agriculture as a producer of potentially negative externalities (Daniels et al. 2010). This type of PES has been able to attract substantial amounts of private sector investment in recent years because the bargain was relatively straightforward and transaction costs in terms of measurement, reporting and verification (MRV) were low. However, there are also major concerns related to such conservation-oriented PES initiatives because they hardly address the challenge of reconciling the environmental objectives of sustainable agriculture with the urgent need to increase agricultural productivity in order to improve global and regional food security.

1.3.1 Opportunities and Constraints of Carbon Markets

PES schemes to avoid negative externalities also received a boost with the creation of the Clean Development Mechanism (CDM) as part of the UN Framework Convention on Climate Change (UN FCCC)'s adoption of the Kyoto Protocol in 1997. The CDM basically paved the way for investments in emission-reduction projects in developing nations. It is the biggest carbon offset market and accounted for over 95 % of total spot and secondary emissions offset trading in 2012.

Originally, the idea of making payments to discourage deforestation and forest degradation was rejected by the CDM because of four fundamental problems: leakage (mere shifting of deforestation activities from protected to unprotected sites), additionality (difficulty of assessing the counterfactual scenario), permanence (risk of reversibility) and measurement (difficulty of measuring the amount of carbon sequestered in trees).

Yet, eventually the role of forests in developing countries as carbon sinks was acknowledged within CDM for the first commitment period. Although deforestation was presented as an important land use change issue, at this stage there were still several unresolved issues associated with Land Use, Land Use Change and Forestry (LULUCF), especially with regard to measuring, reporting and verifying LULUCF activities. During the 11th Conference of the Parties (COP-11) of UN FCCC in Montreal in November 2005, the Governments of Costa Rica and Papua New Guinea as well as the Coalition for Rainforest Nations requested an agenda item on 'Reducing emissions from deforestation in developing countries: approaches to stimulate action' to clarify the objective of Article 2 of the Kyoto Protocol.[11] In Bonn, in 2006, the Subsidiary Body for Scientific and Technological Advice (SBSTA) began considering REDD (Reducing Emissions from Deforestation and Forest Degradation) in developing countries as part of general mitigation efforts aimed at achieving the ultimate objective of the Convention. REDD was officially adopted in 2007 at the UN FCCC meeting (COP-13) in Bali.

In December 2010, at COP-16 in Cancun, REDD formed part of the Cancun Agreements. The Ad Hoc Working Group on Long-term Cooperative Action under the Convention (AWG-LCA) encourages developing country Parties in Paragraph 70 'to contribute to mitigation actions in the forest sector by undertaking the following activities, as deemed appropriate by each Party and in accordance with their respective capabilities and national circumstances:

(a) Reducing emissions from deforestation;
(b) Reducing emissions from forest degradation;
(c) Conservation of forest carbon stocks;
(d) Sustainable management of forest;
(e) Enhancement of forest carbon stocks'.

Points (c), (d) and (e) are generally considered to be the 'plus' of REDDplus because points (a) and (b) have already been acknowledged by the CDM. However, the additional points also triggered controversies that remain unresolved. For example, natural conservation measures are often associated with evictions of indigenous people and history seems to confirm this dark side of ecology (Ankers 2001). In response to this risk, some safeguards have been proposed that respect the

[11] Article 2(ii) of the Kyoto Protocol: 'Protection and enhancement of sinks and reservoirs of greenhouse gases not controlled by the Montreal Protocol, taking into account its commitments under relevant international environmental agreements; promotion of sustainable forest management practices, afforestation and reforestation'.

rights of indigenous people, referring to the adoption of the Declaration on the Rights of Indigenous Peoples. As for the inclusion of sustainable forest management and the enhancement of forest carbon stocks, these points were primarily designed to create additional incentives for industry to buy carbon offsets (FAO 2008, 2011b). The primary concern with this link to carbon markets relates to the high transaction costs linked to MRV of carbon sequestration in land use management. They require large-scale investments to comply with the carbon standards if they are to participate in the global voluntary carbon market. As a consequence, it may become a capital-intensive business that tends to favour large players. The local custodians of the forests, whether they are engaged in forest conservation or climate-smart agriculture (CSA), may still benefit. However, the payments they receive are expected to be small, also because the global financial crisis has increased the uncertainty in the carbon market (FAO 2011b). Worldwide emissions trading volume in 2012 rose by 17 % to 10.3 billion tonnes of carbon dioxide equivalent, compared to the year before, with permits in the EU Emissions Trading Scheme (ETS) accounting for more than three quarters of the total. But, at the same time, prices of EU carbon permits and international offsets fell from 9 to 2.38 euros per tonne in January 2013. The European Commission therefore sought approval from the European Parliament and the European Member States to reduce the number of allowances to be auctioned over the next 3 years (2013–2016) in order to shore up the price. Despite the failure to obtain this approval, the carbon price had somewhat recovered, to 6 euros per tonne, in May 2014. Globally, 39 national and 23 sub-national jurisdictions have implemented or are in the process of implementing carbon pricing, including emissions trading systems.

The World Bank Group and others are encouraging countries, sub-national jurisdictions, and companies to join a growing coalition of first-movers supporting carbon pricing. Emissions trading schemes are currently valued at about US$30 billion, with China now home to the world's second largest carbon market, covering the equivalent of 1115 million tonnes of carbon dioxide emissions.[12] Numerous voluntary carbon markets have also been developed to reward CSA provided that the applied practices meet the requirements of the relevant carbon standard. There are 36 other standard methodologies in this category, which have either been approved or are under development. The most important of these standards developed for credit-generating Nationally Appropriate Mitigation Action (NAMA) is the Verified Carbon Standard (VCR). Compliance with the VCR is a condition for being eligible for payments from the World Bank Biocarbon Fund. It allows projects in the voluntary market to calculate emissions avoided by reducing deforestation either on the edge (frontier) of large cleared areas, like agricultural zones, or in patchwork (mosaic) within standing forests. So far only the Sustainable Agricultural Land Management (SALM) methodology developed within the Kenya Agricultural Carbon Project (KACP) has gained VCS certification (see Chap. 3).

[12] For more information see: http://www.worldbank.org/en/news/feature/2014/05/28/state-trends-report-tracks-global-growth-carbon-pricing

But even in that case, many uncertainties about the carbon market remain, as discussed below (World Bank 2012).

REDD+ would require more private sector investment to cope with the magnitude of challenges. But private sector investment in the aftermath of the global financial crisis and in view of the uncertainty related to the carbon trade is scarce, especially because soil carbon accrues over a long period, while upfront investment costs are often high, making such projects unattractive compared with many other investment options. Government-funded agricultural extension systems may cover large parts of the upfront costs (feasibility studies and pilot phase) and provide an institutional basis for scaling up adoption of good agricultural practices but if returns from the carbon market are expected to be low or at best uncertain, there is little incentive to invest beyond some support from the Corporate Social Responsibility (CSR) side.

MRV of agricultural mitigation activities is difficult and costly. Of the various standards currently used within the carbon market, no single standard meets all of the criteria for broad eligibility of CSA practices. Other uncertainties are related to investor demand, predictable value of credits, reputation for environmental integrity of credits, and responsiveness to local needs. This creates a dilemma for project developers who face trade-offs, notably between the volume and value of credits and the costs and complexities of managing projects.

Problems such as leakage, lack of additionality, and temporality make it problematic to measure the effective benefits of mitigation in small-scale agriculture. For example, it is very difficult for small-scale farmers and pastoralists to reduce deforestation or stocking rates in particular sites without expanding them elsewhere if their livelihood depends on these activities (leakage). Moreover, if the adoption of CSA also makes economic sense for small-scale farmers it becomes difficult to argue that emissions reductions (ER) would not have taken place in the absence of the carbon finance project (additionality). Finally, the permanence of labour-intensive agricultural practices that is required for carbon sequestration may not be assured if the carbon finance project entirely depends on foreign funding (temporality) (FAO 2011b).

All these challenges, however, have not deterred emerging economies such as Mexico, Brazil and China from engaging in massive state-led reforestation programmes that focus not only on carbon sequestration but also on the prevention of soil erosion, the provision of watershed services and the enhancement of biodiversity.

1.3.2 Biodiversity Conservation Markets

The continuous loss of biodiversity, especially in the hotspot areas of developing countries, is a great concern in the international community. In view of the need for a global assessment of the state of biodiversity and ecosystem services and improved conservation policies, the Intergovernmental Platform on Biodiversity and Ecosystem Services (IPBES) was established in April 2012. It will complement and coordinate the numerous existing public and private initiatives to protect biodiversity in situ and ex situ and make use of policy instruments that encourage the creation of markets for biodiversity.

In the broadest sense, biodiversity markets include any payment for the protection, restoration, or management of biodiversity (e.g., biodiversity offsets, conservation easements, certified biodiversity-friendly products and services, bioprospecting, payments for biodiversity management, hunting permits, and ecotourism). Such markets can generally be based on government-mediated payments or on private contributions to voluntary markets comprising certified biodiversity-friendly products, donations for biodiversity conservation or research, positive public relations, ecotourism and recreation, and others (Madsen et al. 2010).

The creation of a biodiversity market that would facilitate trade in biodiversity offsets as a commodity (by analogy to carbon offsets in the carbon market) is one of the most prominent ways of slowing down biodiversity loss in situ in developing countries. Biodiversity offsets, defined as measurable conservation outcomes of actions, are designed to compensate landowners in developing countries for avoiding significant residual adverse impacts on biodiversity arising out of their particular land use activities. As illustrated by the US policy, such markets for biodiversity and wetland mitigation often emerge in response to environmental regulation. The main purpose of trading biodiversity offsets is to achieve no net loss and preferably a net gain of biodiversity on the ground with respect to species composition, habitat structure, ecosystem function and people's use and cultural values associated with biodiversity. In addition, conservation finance initiatives are exploring innovative ways to raise investment for biodiversity conservation to close the gap between the combined public funding for biodiversity protection and the estimated funding required to halt the loss of biodiversity in aquatic and terrestrial ecosystems. However, the challenge of monetarizing the value of biodiversity remains considerable, in addition to the problems already discussed in the previous section on carbon projects (leakage, additionality, temporality). The Vibrant Oceans initiative was started in January 2014 thanks to a US$50 million investment by former New York Mayor Michael Bloomberg. The initiative aims to overcome these constraints by promoting sustainable small-scale and industrial fishing practices (including advocating industrial fishing reforms such as catch limits) designed to prevent the degradation of aquatic ecosystems and, at the same time, boost fish populations in Brazil, Chile, and the Philippines—which account for 7 % of the world's fisheries. EKO Asset Management Partners, another partner in the initiative, will develop investment blueprints to attract private capital and financially reward local fishermen and industrial fleets that successfully manage the transition to sustainable fishing practices. If the initiative succeeds, it could serve as a model for future global reform efforts.[13]

[13] For more information see: http://www.theatlantic.com/business/archive/2014/02/bloombergs-new-plan-save-the-ocean-feed-the-world/283546/

1.3.3 From a Project-Based Approach to a Landscape Approach in Terrestrial Biodiversity Conservation

Even though the experience in industrialized countries is that cap-and-trade regulated biodiversity offset schemes can indeed create real markets and be very powerful when used correctly, such markets also face considerable criticism and have even led to campaigns against biodiversity offsets.[14] This is because there are two main problems with the offset schemes.

The first problem is that biodiversity offset frameworks may not be implemented according to the mitigation hierarchy (avoiding and minimizing impacts before proceeding to compensatory mitigation) and therefore lead to a 'licence to trash' meaning that development takes place in areas where impacts should have been avoided or more effectively minimized. The second problem with biodiversity offsets relates to the unclear issues of equivalence and location. Should offsets be as close to the original site as possible so that the local population benefits from the compensation scheme? Or should they be in a remote site where the value of biodiversity is more equivalent to the original site. Policy guidance is not very clear on this aspect.

Some authors therefore propose to move from a narrow project-based approach to a broader landscape approach that would require landscape plans. Such plans would look at the cumulative impacts of current and projected development, and, based on the findings, identify the mitigation hierarchy that should be applied (i.e., avoidance versus offsets). This would ensure that mitigation is in accordance with broader conservation goals to maintain large, resilient ecosystems that support biodiversity habitats with sufficient connectivity and benefit human communities (McKenney and Kiesecker 2010).

A biodiversity offset framework that takes into account landscape planning, however, requires strong government oversight, effective legal systems, enforcement of rules and regulations, and robust financial institutions. These conditions may be met in some industrialized countries, but they are absent in most parts of the developing world (Worldwatch Institute 2008). Even though, in places such as South Africa, Colombia, Mexico and Brazil, laws requiring or encouraging biodiversity offsets are either being considered or are already being implemented, the instruments to effectively enforce them are often lacking. The only successful enforcement of rules pertaining to biodiversity conservation has been observed in China and there it was not linked to mitigation banking.

[14] See also http://naturenotforsale.org/

1.3.4 China's Approach to Ecosystem Services Management

The Chinese government's programme 'Grain for Green' (the official title translates as the Sloping Lands Conversion Programme, or SLCP) pays farmers to keep forest cover on hillsides. This programme was part of a policy response to the severe Yellow River drought in 1997 and the extreme floods along the middle and upper reaches of the Yangtze River in 1998. A commission set up by the Chinese Academy of Sciences identified the erosion of land cleared for farming as one of the major contributors to the devastating floods, which caused damages that amounted to US$30 billion, displaced 18,000 people and led to thousands of lives being lost (Reardon 2012). The Chinese government subsequently initiated the SLCP in 1999. Its aim was to help conserve watersheds and prevent floods, but it was also related to biodiversity conservation. The money to pay for the environmental services comes directly from tax revenues and its redistribution is based on certain established criteria. While the SLCP system does help increase the value of standing forests, it does not directly link the users of the biodiversity services with the providers of those services as classic PES would suggest. The government acts instead as a mediator in the transaction and determines the amount that farmers are paid to take part in the reforestation scheme (WRI 2008). 120 farmers had obtained a payment by 2008, and a study in 2011 showed that the vast majority of them have used the payment to leave farming and find employment elsewhere (Li et al. 2012). This highlights the importance of economic growth outside agriculture in addressing the social dimension of ecosystem management schemes that focus on the avoidance of negative externalities. With this policy, China has managed to add 1.6 % to its forest cover over the past decade. During this period the Chinese government has invested more than US$100 billion in what it calls 'ecocompensation' schemes (Reardon 2012). Even though these numbers are impressive, it is not clear to what extent the scheme has resulted in leakage abroad (e.g., Chinese companies that shift their unsustainable logging practices to southeast Asian countries).

At any rate, the swift and reasonably effective response of the Chinese government response indicates that a strong legal and financial effort to introduce, conserve or restore valuable ecosystems in a developing country may have to be linked to an event that creates a general awareness of the economic importance of environmental services. The conservation of biodiversity may then be a welcome side effect, but it can hardly be the driver of investment in environmental services in less developed countries that still need technological and economic change to become sustainable and therefore cannot afford to focus on conservation alone. This was also the case in Europe in the nineteenth century. The environmental damage caused by the rapid industrialization process led to the first forest laws, which were designed to ensure that natural resources were used in a sustainable way. As in China today, it was off-farm employment as well as technological change in agriculture that helped to enforce these forestry laws without causing an increase in hunger and poverty (Aerni 2009).

1.3.5 *Ecosystem Management in Pastoralist Systems*

Pastoral systems, involving livestock farming or grazing, characterize two-thirds of global dryland areas. These ecologically sensitive areas are highly prone to desertification and are often characterized by poverty and economic stagnation. It is estimated that one billion people depend on livestock for their livelihood, and livestock serves as at least a partial source of income and food security for 70 % of the world's 880 million rural poor. Degradation of the land base not only negatively affects livestock productivity but also the accumulation of carbon in the soils. Improved pasture and rangeland management and wildlife conservation in dryland areas could help restore the soil carbon sink, improve soil health and biodiversity, reduce risks of drought and flooding and improve the livelihood of pastoralists. In this context, payments could either provide incentives for sustainable livestock management practices (asset-building) or for protecting wildlife from the effects of expanding pastoralist activities while also ensuring alternative revenues for pastoralists (use-restricting) (FAO 2010).

1.3.6 *Sustainable Livestock Management*

PES can potentially play a significant role in addressing trade-offs between livestock production and environmental services. Livestock production represents an essential provisioning service in developing countries. It provides not only nutritious food, but many other useful products, especially in rural households, such as material for clothes and shelter, energy (biogas), manure, means of transportation, and capital stock in uncertain times (Abdulai and CroleRees 2001). On the other hand, livestock production creates negative environmental externalities, especially when it is accompanied by strong population growth and little access to the technology that would allow for more efficient use of the scarce resources (FAO 2012). These environmental externalities are particularly prominent when livestock keepers strive to optimize the direct benefits from livestock, such as income, in the absence of regulations or incentives to mitigate livestock's negative impacts on the environment.

1.3.6.1 Matching Grants

One way of addressing these negative externalities is through matching grants.[15] Matching grants have increasingly been used in the context of international

[15] The concept of a matching grant is simple; an institution designates funds to go to particular types of projects. Groups within the community can then develop project proposals and apply for the grant. If accepted, the institution will match the community contribution to the project.

development by multilateral and bilateral institutions, including the International Fund for Agricultural Development IFAD and the World Bank. Although initially confined to public goods investments, they are being used more and more often to finance productive assets and investments by communities, groups and individuals that are expected to have positive externalities for the environment in the region. However, matching grants entail substantial risks. Even though the intention of the World Bank and others funding bodies is to apply matching grants as an incentive for sustainable grassland management, such schemes must be carefully designed to avoid crowding out private and public investments, as well as elite capture and rent-seeking behaviour, leading to poor use of scarce public funds for the benefit of the few. To ensure that the matching grants are correctly used in accordance with project objectives, there should be an exclusion list detailing activities that the project will not support through the matching grant facilities. This list would include activities that are not seen to be financially or socially responsible, or which the individuals and/or communities could be expected to self-finance. Second, there should be careful screening of grant applicants with respect to group registration, constitutions, management committees, business plans and auditability. Third, activities which are eligible for funding from other sources, or which could be financed by existing financial service providers would not be eligible for matching grants, although the project could provide assistance in accessing these forms of finance. Fourth, all grants would be disbursed in tranches, with each payment being conditional upon independent verification that the funds were being used by the intended beneficiaries in accordance with the grant agreement.[16] However, there are many indications that matching grants may help jumpstart innovation and entrepreneurship in the cost-effective provision of environmental services—if criteria for access to such grants are not too exclusive and are designed to reward innovation. This would increase the potential for local people not only to benefit from improved provision of environmental services but also from productivity increases as well as new markets and new jobs. Local entrepreneurs who aim to access, tailor and commercialize innovation with positive environmental effects should certainly assume part of the risk by providing a share of the funding. To enhance the capacity to do so, voucher schemes could be introduced or local banks and private sector networks could be created with a focus on the creation of environmental markets. This approach is working well in Rwanda (Wongtschowski et al. 2013).

[16] For more information see: http://www.ifad.org/evaluation/public_html/eksyst/doc/insight/pa/ghana.htm

1.3.7 Payments for Wildlife Conservation

The PES schemes that have proved to be workable for all parties were mainly designed to avoid the negative externalities produced by the different forms of livestock systems (pastoral systems, mixed extensive systems, mixed intensive systems, intensive livestock production). Yet the financial sustainability of most of these use-restricting projects is far from guaranteed owing to numerous problems and high transaction costs related to the MRV of PES. There is also uncertainty about the additionality criteria that can be compromised by leakages. 'Leakage' arises when the conservation problem being addressed by PES is shifted elsewhere, for instance, when a landowner who receives payment for not moving his herd into a protected area simply increases the grazing activity elsewhere causing the environmental problem on another piece of land that is not subject to the contract (Meijerink 2008).

There are however examples of situations in which these problems were effectively addressed, transaction costs for MRV could be held sufficiently low and financial sustainability seems to be ensured thanks to the active interest of the private sector stakeholders paying for the service (avoidance of the negative externality).

The Wildlife Lease Program (WLP) in Kenya effectively addresses these concerns. It involves direct monetary payments to pastoral Maasai landowners living to the south of Nairobi National Park (NNP). The Payment for Wild Life Conservation (PWC) scheme was started in 2000 to maintain the seasonal wildlife dispersal areas and migration corridor in Kitengela to ensure the viability of the Nairobi National Park (NNP) ecosystem and its biodiversity; and to enhance the economic security and quality of life of local pastoral landowner households (Gichohi 2003). The scheme is coordinated by The Wildlife Foundation (TWF), a local NGO which acts as an intermediary between pastoral land users and ecosystem services buyers. The scheme started in 2000 with 86.6 ha of land provided by two landowners. In 2010, 375 households had enrolled in the scheme, providing a total of 16,500 ha of land. Funding comes from voluntary national and international buyers. The main funders are the Kenya Wildlife Service (KWS), the state agency responsible for wildlife management, the Global Environment Facility (GEF) and The Nature Conservancy (TNC). Participating households sign a 1-year lease, and are paid US$10/ha per year (with a 5 % base annual inflation factor), an amount that is competitive with returns from livestock grazing, which is not precluded as a land use for participants. Pastoral landowners have a double benefit: they receive an income from PES payments and are still able to continue with pastoral livestock production. In return, land users participating in the programme are required to allow free movement of wildlife on their land, refrain from poaching wildlife, report poaching by others, protect natural vegetation on their land, and avoid fencing or sub-dividing their land (Silvestri et al. 2012).

Encouraging farmers to move away from unsustainable livestock practices and instead to adopt sustainable silvopastoral management systems may be considered as an example of how PES in livestock is not just about paying for the reduction of negative externalities of livestock management but also for the production of positive externalities (growing trees to sequester carbon, improved livelihoods).[17] But this represents a clear shift away from livestock towards agroforestry as the Regional Integrated Silvopastoral Ecosystem Management Project (RISEMP) illustrates. RISEMP has been implemented in Colombia, Costa Rica and Nicaragua, in areas that were seriously degraded as a result of unsustainable livestock farming. The goal was to restore these areas by protecting the soils, storing carbon, and fostering biodiversity. For that purpose, PES were designed in a way that would make it beneficial for farmers and communities to halt unsustainable practices and participate in the efforts to restore the degraded land (Porras et al. 2008).

International demand for carbon and biodiversity benefits came from the GEF, World Bank and the FAO. From 2003 to 2006, cattle farmers received between US $2000 and US$2400/year/farm, representing 10–15 % of net income to implement the silvopastoral systems programme. During the initial 4 years, the high investment costs of implementing the silvopastoral scheme reduced the economic benefits to the farmers (for example, because of the costs of live fencing and of planting trees). Once the trees had grown sufficiently to provide benefits, as a result of the greater availability of high quality fodder and shade from the trees, milk production was expected to increase and expenditure on fertilizers and pesticides, as well as irrigation was expected to fall. PES was designed in this project in such a way as to raise the profitability of the silvopastoral systems up to the level of the traditional ones by the 2nd year and to increase profits by 50 % by the 5th year. The project resulted in a 60 % reduction in degraded pastures in the three participating countries, and the area of silvopastoral land use (e.g., improved pastures with high density trees, fodder banks and live fences) increased significantly. The environmental benefits associated with the project included a 71 % increase in carbon sequestered (from 27.7 Mt of CO2eq in 2003 to 47.6 Mt in 2006), increases in bird, bat and butterfly species and a moderate increase in forested area. Milk production and farm income also increased, by more than 10 % and 115 %, respectively. Herbicide use dropped by 60 %, and the practice of using fire to manage pasture is now less frequent. Other demonstrated environmental benefits of silvopastoral systems included improvements in water infiltration, soil retention, soil productivity and land rehabilitation, as well as a reduction of dependence on fossil fuel (e.g., substitution of inorganic fertilizer with nitrogen-fixing plants) (Porras et al. 2008). Nevertheless, silvopastoral systems face many challenges related to the appropriate institutional design that are still being addressed (Pagiola and Arcenas 2013). The major challenge that is not discussed in existing evaluations is financial sustainability. Land users are confronted with a significant increase in labour costs and time required to maintain such systems and it remains doubtful whether they could continue without the support of GEF and the World Bank. The challenge of

[17] For a detailed account of PES in wildlife conservation schemes in Kenya see also http://www.fao.org/fileadmin/user_upload/pes-project/docs/FAO_RPE-PES_OOConservacy_Kenya.pdf

avoiding reversibility is great, especially when the focus is on avoiding negative externalities.

In addition to matching grants, wildlife conservation schemes and silvopastoral ecosystem management, various pilot projects apply PES to livestock management. Examples include livestock inclusive agricultural production systems (LiAPS) that have proved successful and scalable. Yet, such projects are still rare, especially with regard to the remuneration or reward for positive externalities (e.g., animals play an important role, not just as a capital stock but also in harvesting and relocating nutrients, increasing the supply of manure, making the best use of crop residues (used as feed, incorporating the remaining stalks directly into the soil, where, for some time, they act as a nitrogen trap) (ILRI 2012).

1.4 The Problem with PES Schemes That Aim at Avoiding Negative Externalities

In many respects, the concern for the environment is a salient political issue in industrialized countries where the public increasingly questions the expected gains from economic and technological change in the face of expected losses related to worry about the decline of cultural diversity and environmental quality. It is well known that population growth without any means to increase ordinary people's income prior to the Industrial Revolution in Europe had serious social consequences (poverty, workers' health, inequality) and environmental impacts (deforestation, pollution from waste products). Social justice was a particularly salient issue at that time because of the European reliance on food imported from plantations overseas that depended on slave labour and the exploitation of peasants in the manorial agricultural system that prevailed in Europe (Aerni 2011a). Governments back then responded to these social and environmental challenges by investing in new economic opportunities and technological change rather than preventing them.

1.4.1 Fundamental Attribution Error in Affluent Societies

In view of the progressive response to sustainability challenges in the nineteenth century, it may be a fundamental attribution error (Harman 1999) in today's affluent societies to blame technological change and economic development for environmental problems in developing countries. This fundamental attribution error may explain why many developing countries tend to oppose the prevailing conservationist view in donor countries and prefer a progressive approach to sustainable environmental management (Juma 2011). Preserving the status quo would be especially unsustainable in LDCs. They need structural change to become sustainable because they are facing huge challenges: coping with strong population growth, lack of private sector investment, youth unemployment, lack of access to technology and knowledge, massive environmental degradation as a result of deforestation, overuse of natural resources, and soil erosion. Unlike in developed countries where affluence is considered to be the main threat to sustainable development, it is poverty that may pose the biggest sustainability problems in LDCs (Gilbert 2012; Beckerman 2002).

Developing countries that are currently at the beginning or in the middle of an industrial transformation process (comparable to the industrial transformation taking place in Europe in the nineteenth century), still exhibit many of the traditional socioeconomic structures and agro-environmental conditions that have disappeared in affluent countries. Yet the assumption of the public in affluent countries that these traditional institutions represent a social value that needs to be defended may be misleading because such institutions might have functioned in a sustainable manner as long as average life expectancy was very low and child mortality high—ensuring low population growth rates. However, once the population has started to grow, as a result of increased access to medicine and food, clinging to traditional institutions may become part of the problem rather than part of the solution and the pressure for change comes from within rather than from outside the community (Boserup 2005). The strengthening of traditional institutions also results in the strengthening of the traditional elite that does not want change because it benefits from the status quo. This has led to unholy alliances throughout history, between the white European elite and the African elite (Ankers 2001; Mandela 1995; Appiah 2007; Aerni 2009, 2005). Many representatives of the South have therefore argued that the implementation of western-funded conservation schemes (paying landowners for avoiding negative externalities) may go hand in hand with the erosion or loss of rights and the exclusion from lands, territories and resources of indigenous people that largely rely on communal rights (Leach and Scoones 2013; Fairhead et al. 2013). Safeguards have therefore been proposed with regard to all sorts of nature conservation schemes that would protect the rights of indigenous people. It remains uncertain, however, how the protection of these rights could be enforced (Moss and Nussbaum 2011).

1.5 A Shift Towards the Remuneration of Positive Externalities

Even though PES schemes to avoid negative externalities tend to have lower transaction costs and attract more private sector investment (Wunder and Boerner 2011), there is also a trend towards PES schemes to remunerate farmers for the generation of positive externalities in agriculture, especially with regard to watershed management projects in mountain regions in the developing world. This trend has been initiated by international environmental reports that embraced the European concept of multifunctionality by recognizing the positive role farmers play in the provision of environmental services (MEA 2005; FAO 2007; Padulosi et al. 2012). Moreover the global food crisis, that has been more or less acute since 2008, made many stakeholders aware that environmental conservation should not come at the expense of a decrease in agricultural productivity.

Finally, the UN Environment Summit in Rio in 2012 (Rio+20) concluded that the food crisis and the environmental crises have to be jointly addressed by promoting a Green Economy. This implies that farmers should not just receive support to maintain ecosystem services but also to increase productivity by means of sustainable intensification.[18]

1.5.1 UN Millennium Ecosystem Assessment as a Promoter of RPE

The idea to remunerate farmers for the creation of positive externalities (food security, climate change mitigation, landscape beauty) gained prominence in developing countries in response to the UN MEA report in 2005 and its call for agricultural practices that support the provision of ecosystem support services (MEA 2005). According to the report, such supporting services comprise provision services (e.g., food and fresh water), regulating services (such as water purification and climate regulation), and cultural services (landscapes of aesthetic, spiritual and cultural importance). The report suggests that enhancing provisioning services, such as food production, over the past 50 years would have been achieved at a high price in the form of the degradation of many ecosystem services, increased risks of nonlinear changes, and the exacerbation of poverty among some groups of the population. In this context, the authors of the report argue that the 'nonmarketed' benefits (non-trade concerns) related to ecosystem services have to be taken into account in order to discourage further damage to ecosystem support and regulating services. In this context, PES were considered to be one important tool for taking such benefits into account. Even though the report admits that the use of new technologies in agriculture and new methods of ecological restoration are essential in coping with future environmental challenges, these activities are not really portrayed as positive externalities that are generated by the private sector through investment in technological innovation. Instead, the recommended policy instruments mostly focus on the regulation of the potentially negative externalities of technological and economic change in agriculture. This bias may be related to the implicit baseline assumption of the report that most of the degraded natural environment today is a result of the expansion of industrial agriculture in the twentieth century.

The report had a significant impact on the international discourse on agriculture, food security and the environment. While private-sector driven modernization of

[18] See http://www.fao.org/agriculture/crops/core-themes/theme/spi/en/

agriculture was increasingly considered to be a source of negative externalities that need to be strictly regulated, farmers that are assisted by NGOs and the public sector are to be compensated for the adoption of sustainable agricultural practices that generate positive externalities ('nonmarketed' benefits) for the environment, culture, society and human nutrition. This view may be too simplistic to cope with the complex double challenge of mitigating climate change and ensuring global food security.

1.6 The Challenge of Financial Sustainability

Public–private partnerships in developed (Perrot-Maître 2008) and developing countries (Willy et al. 2012) are producing an increasing amount of positive externalities for society and the environment. Rather than being part of the problem, the private sector can be part of a sustainable solution, not just because of its financial contributions but because of its know-how about the development and sustainable use of technology and its professional expertise in business administration. All these assets complement those of NGOs (including community organizations, trust-networks, awareness building, etc.) and ensure financial sustainability by converting a PES scheme into a local market for innovative environmental goods. There is even evidence that these public–private partnerships are compatible with the goals of social empowerment of indigenous people.[19]

Whereas the remuneration of these positive externalities in agriculture is largely the responsibility of the state in developed countries, developing countries that have adopted legislation to support the adoption of sustainable agricultural practices must often rely heavily on foreign donors to fund compensation schemes, including PES, to remunerate farmers for the positive externalities they generate. The voluntary nature of a PES agreement consisting of a seller (farmer), an intermediary (donor or driver) and a buyer (the contributing beneficiaries of the services in the private sector) makes it difficult for donors (often the main driving force behind PES) to exit without putting the future of the PES scheme at risk. After all, their financial support as intermediaries of the PES scheme is often much greater than the contributions from the local buyers. Moreover, the valuable assistance they offer to the farmers that adopt sustainable agricultural practices and thus become sellers of the environmental services would also need to continue.[20] The financial sustainability of a PES project cannot therefore be ensured unless a PES scheme also

[19] The case of the Maoris and their efforts make more sustainable use of their forests by means of a public–private partnership is illustrated in the following factsheet: http://www.fao.org/fileadmin/user_upload/pes-project/docs/FAO_RPE-PES_NgatiP_NewZealand.pdf. And the case of the Maasais generating revenues from an abattoir as well as its waste products is illustrated at: http://www.fao.org/fileadmin/user_upload/pes-project/docs/FAO_RPE-PES_Kiserian_Kenya.pdf

[20] Despite many promising developments in terms of financial sustainability, the Lake Naivasha scheme in Kenya to improve water quality through a PES Watershed project continues to be

generates other benefits such as increases in agricultural productivity, new revenue-generating innovative technologies and practices, and business opportunities emerging from sustainable livelihood projects. Such additional revenues generated through the creation of new local markets for environmental goods offer a viable exit strategy for the initial sponsors (IFAD 2009). Such markets will not emerge unless the private sector has more incentives to invest in sustainable practices in land use management.

1.7 The Role of Private Standards and Product Certification

The need to make PES schemes more financially sustainable must also be seen in the context of the growing trend in the agriculture and food business towards private standards and recognition certificates in developed as well as developing countries, aimed to ensure sustainable agricultural practices, food safety, food quality, and worker welfare. Since private standards are often combined with RPE/PES schemes, they can also be considered as part of an evolving ecosystem market. Farmers are offered a premium price in return for complying with particular sustainability standards.

A basic distinction must be made, however, between global private standards designed exclusively by global retailers for large agricultural exporters, and standards that are negotiated locally between a direct buyer of the output (or a direct farm input provider) and the farmers who have adopted sustainable agricultural practices. Since global retail standards are primarily designed to please consumers in affluent countries and not necessarily farmers in poor countries there is very little flexibility to adjust such standards to particular local circumstances. It is therefore not certain to what extent such private standards actually make it more difficult for local small-scale producers to participate in international trade (Freidberg 2010; Aerni 2013; Cramer et al. 2014).

Global private standards have become a de-facto condition for developing countries (especially in the area of food safety) wishing to export to more affluent markets. However, if such private standards become very costly to comply with they may lead to a concentration in agroindustry and the abandonment of outgrower schemes (small-scale farmers that obtain the input to grow export crops from a large agricultural exporter and then sell the produce back to them for export) (Freidberg 2007). Private standards can nevertheless have a very positive impact on the sustainability record of the agricultural export business. In the case of the flower business in Kenya, the need to comply with private standards led to a new market

strongly dependent on outside support: http://www.fao.org/fileadmin/user_upload/pes-project/docs/FAO_RPE-PES_WWF-Kenya.pdf

for wetland mitigation systems because each certified flower farm must have its own wetlands to clean waste water. This will be further discussed in Sect. 3.3.

1.7.1 *Locally Designed Certification Schemes*

The objective of locally designed and applied certification schemes is different. Such schemes are more responsive to the local agro-ecological context, take farmers' knowledge into account and may facilitate markets for environmental goods based on local trust networks. Farmers who have adopted certain locally appropriate sustainability measures that produce positive externalities for society and the environment in the region (e.g., less water pollution, more soil conservation, greater biodiversity) may obtain a certificate. This certificate allows them either to sell their agricultural produce directly for a price premium to a local buyer (e.g., large employers such as universities, international companies or hospitals) or to buy input from the local cooperative or agro-dealer for a specified discount (Aerni 2013). Through such schemes, new social networks can be created between marginal rural regions and urban centres. This in turn can result in further ties that create additional business opportunities for small-scale farmers (e.g., regular buyers that obtain their products from a particular farm might also encourage their employees, students or patients to visit or spend a holiday on the farm and thus jumpstart a small agro-tourism business). An important condition for such schemes to work is a corps of public or private extension workers that are in a position to monitor, report and verify the sustainable agricultural practices that a farmer claims to have adopted. Such locally designed certification programmes to make PES schemes financially sustainable by creating new local markets for environmental goods are believed to have a great potential to ensure the financial sustainability of PES (see also the Lake Naivasha case study in Sect. 3.3).

1.8 Conclusions from a Historical Point of View

Most international organizations, including the World Bank, UNCTAD, OECD, UNIDO, WTO and FAO, share the view that prevailed at the Rio+20 UN Summit that there needs to be a global transformation towards a greener economy. The concept of the Green Economy emphasizes the crucial point that economic growth and environmental stewardship can be complementary strategies, challenging the still commonly held view that the two objectives are mutually exclusive—in other words, that the synergies prevail over the trade-offs (UN-DESA, UNEP, UNCTAD 2012). This chapter reviewed the history of PES and RPE in agriculture and basically reached the same conclusion: efforts to ensure the provision of ecosystem services cannot just rely on the sellers and buyers of ecosystem services but require investments in innovation and the creation of markets for environmental goods in order to become financially sustainable. Many concrete cases in developed and developing countries have been used in this chapter to illustrate the value of entrepreneurship and innovation in the sustainable provision of environmental services. Yet, these aspects have so far not been addressed in the academic literature of PES, as the following chapter will show.

Chapter 2
Payments for Environmental Services: Revisiting the Theoretical Baseline Assumptions

Abstract The concept of payments for environmental services (PES) has its theoretical roots in neoclassical welfare economics. The concept suggests that the degradation of environmental resources is linked to the fact that these resources are considered to be for free. By assigning a monetary value to environmental services, sufficient incentives for market players would be created to protect, trade and invest in the provision of environmental services. The implicit assumption that once you assign such a value, a market would automatically evolve with buyers and sellers of the environmental service does however hardly work in practice because it is based on a comparative-static rather than a dynamic understanding of sustainability. This chapter illustrates that a flourishing market of environmental goods and services cannot be merely designed and funded by an external agent. It requires instead active local entrepreneurs that generate revenues through innovation.

The rise of the concept of payments for environmental services (PES) is not just the result of a reframing of agricultural, development and environmental policies after the Cold War, but also reflects the increasing use of policy instruments that have their theoretical roots in neoclassical welfare economics. The basic idea behind payments for ecosystem services and the remuneration of positive externalities is that markets tend to ignore the value of environmental services and thus contribute to overuse and eventual depletion of natural resources. Neoclassical economists argued that this so-called 'tragedy of the commons' results in the destruction of common pool resources. It could be averted, so they argue, by assigning a monetary value to environmental services. This would provide sufficient incentives for market players to protect, trade and invest in the provision of environmental services.

But are these theoretical assumptions actually in line with the practical outcomes of diverse PES schemes in developed and developing countries? Evidence from the history of PES (Chap. 1) and the field visits to Kenya (to be discussed in Chap. 3) suggest they are not.

Nevertheless, PES continues to be considered an essential policy instrument to implement several multilateral environmental agreements (MEAs) that have been ratified over the past three decades to protect valuable ecosystem services and promote the sustainable use of natural resources. The number of operational Watershed PES programs alone almost doubled from 2011 (US$8.2 billion for 203 projects) to 2013 (US$12.3 billion for 402 projects (Bennett and Carroll 2014).

P. Aerni, *The Sustainable Provision of Environmental Services*, CSR, Sustainability, Ethics & Governance, DOI 10.1007/978-3-319-19345-8_2

In this context, national and international financing mechanisms have been created to fund projects that aim at the sustainable management of ecosystem services. Since 1991, the Global Environment Facility (GEF) alone has disbursed US$10.5 billion in grants and leveraged US$51 billion in co-financing for more than 2700 projects in over 165 countries.[1] Of these, the GEF has funded 42 projects where PES was a core element or part of the project design. Investments of US$70 million have been made in 14 projects where PES is central to the project's design. With this amount it was possible to leverage an additional US$395 million in co-financing. In addition, 28 projects of which PES is a part or a minor element have been supported by the GEF with an amount of US$152 million (leveraged to US$518 million).

Even though the system of global environmental governance has generated new treaties, more money and more participatory approaches to improve global environmental management, preserve natural capital and prevent climate change, the global environmental challenges continue to increase and pose a global threat to humankind (Schiermeier 2012; Najam et al. 2006; MEA 2005). There is therefore a growing concern that many environmental policy instruments that have been applied for more than three decades and that continue to shape global and national environmental regulation, may have to be revisited from a theoretical and practical point of view and, in some cases, redesigned to ensure their effectiveness in addressing the global environmental challenges of the twenty-first century. The concept of PES is one of the instruments where the straightforward theoretical argument to pay land users for the provision of environmental services is often confronted with a more complex reality.

The following section reviews the evolution of the theory and practice of PES, highlights certain inconsistencies between theoretical claims and experience on the ground, and illustrates how the current trend towards hybrid PES projects copes with the practical challenges. Hybrid PES no longer follows the strict theory on PES derived from neoclassical welfare economics; instead its purpose is to combine effective policy instruments on different policy levels with new business opportunities linked to the provision of environmental services. Such hybrid PES projects may still be based on the idea of rewarding the contributions of farmers to sustainable landscape management, but they are generally more open towards innovation and encourage local entrepreneurship. As a result, the private interest of local landowners in increasing their household revenues becomes more closely aligned with the public interest in managing ecosystem services sustainably. This ultimately ensures the financial sustainability of a PES project and thus the permanence of environmental improvements achieved during the external funding phase of the project. Classical PES theory could not have anticipated this trend. This chapter therefore identifies a need to revisit the theoretical baseline assumptions of welfare economics that still guide most PES projects. The basic principles of welfare economics were developed during the Cold War period and continue to

[1] For more information see: http://www.thegef.org/gef/sites/thegef.org/files/publication/PES_english.pdf

be widely applied in today's global knowledge economy. These principles tend to ignore the role of innovation and entrepreneurship in enabling knowledge generation and its potential to be effectively used to address environment and development challenges.

If governments provide the necessary enabling environment, innovation may lead to the creation of new goods and services that produce considerable social and environmental welfare benefits. Governments therefore have a responsibility not only to regulate potentially unsustainable change but also to facilitate sustainable change through an incentive system that rewards the search for new solutions to environmental challenges. In this context, PES may indeed assume the role of a vehicle to create new markets for environmental goods, but it is not a business case by itself, as many neoclassical economists tend to assume.

2.1 Introduction

2.1.1 The Evolution of Environmental Policy Tools

The emergence of the modern environmental movement in the 1960s created the new field of environmental and resource economics (Turner et al. 1994) and resulted in the first institutions of environmental governance on the national level (US Environmental Protection Agency in 1970), the international level (United Nations Conference on the Human Environment in Stockholm 1972) and the supranational level (when the EU created the Environmental and Consumer Protection Directorate in 1973).

The policy tools for implementing environmental regulation are designed either to restrict market functions by means of regulation of access to resources (direct controls via standards, bans and technical specifications), to secure the conservation of natural capital and environmental services, or to bring market efficiency to the supply of environmental goods (price incentives via taxes, subsidies, tradable permits and PES). In addition, there are policy tools related to moral persuasion (voluntary compliance) and the public production of environmental goods (e.g., large-scale waste-water treatment) (Oates and Baumol 1975; van Noordwijk et al. 2012).

Policies that address a particular environmental problem, however, are mostly based on hybrid policy programmes that combine price incentives with direct control of the implementation process (Oates and Baumol 1975). This has led to a wide variety of combinations of environmental policy instruments in different countries. These reflect not just the particular financial endowment of a country for protecting the environment and the cost-effectiveness of policy tools derived from the insights of ex-post evaluations, but also the agendas of political stakeholders and how they shape the public perception of the presumed causes of environmental problems and how to address them (Aerni and Bernauer 2006). What all these policies had in common was the claim that they were based on widely accepted legal principles of environmental law. These principles comprised the principle of proportionality, the precautionary principle and the 'polluter pays' principle. All of these principles, however, are very vague and leave plenty of policy space for interpretation and application (Malanczuk 1997). The underlying theoretical baseline assumption of these principles, as well as of the policy tools that have been derived from them, is that economic development causes negative externalities for the environment that must be internalized either by restricting access or by assigning a value to environmental resources and thus providing incentives for decision makers to take the environmental costs of their actions into account (Costanza et al. 1997). In many cases applying these principles and baseline assumptions definitely helped to improve the provision of environmental services; but the implicit negative framing of economic and technological change (perceived as a threat rather than as part of the solution to the environmental challenges) may do a disservice to the long-term provision of environmental services because sustainable environmental management practices that are commercially viable (ensuring that the environmental improvements are not reversed once external subsidies end) must be private-sector driven. There is no question that the political rhetoric on how to save the environment from man-made change (Jonas 1985; Beck 1992; Pollan 2006; Ehrlich and Ehrlich 2009) has shaped the public debate. In this context, many stakeholders in politics attempt to gain public trust by portraying themselves as concerned citizens who are attempting to halt the actors of economic and technological change through protest action (Aerni and Bernauer 2006).[2] This negative framing is not just undermining private sector investment in new technologies that have the potential to substitute for existing unsustainable products and practices but has also led to a general neglect of public investment in R&D to facilitate sustainable change through innovation rather than merely regulating possibly unsustainable change (Prins and Rayner 2007; Juma 2011).

2.1.2 PES: Is There a Gap Between Theory and Practice?

Markets create prices that are meant to reflect the actual demand and supply of the goods and services on offer. The price signals to market participants whether there is under- or over-provision of a particular good or service. This again influences the allocation of scarce resources to the production of that particular good. Yet, preferences expressed on the demand side (consumers) are often not exogenous (i.e.,

[2] This does not mean that there would be no genuine problems resulting from global change that needed to be urgently addressed and no citizens that really have the public interest at heart. But there is also opportunism in movements of resistance often resulting in unholy alliances between those who benefit from the status quo (e.g., subsidized farmers or a subsidized coal industry) and advocacy groups that need support for their radical opposition to technological and economic change.

based on a rational preference ranking of fixed possible choices constrained by certain budget restrictions) but endogenous (shaped by marketing, political framing, persuasion and public opinion) (Nisbet 2015). So the ideal market situation almost never applies to real markets that are shaped by human beings and their emotions, beliefs and imperfect knowledge (Lichtenstein and Slovic 2006). Moreover, markets often fail to signal genuine scarcity, which may lead to the depletion of common pool resources (e.g., potable water, fertile soil, clean air) that are non-rival and non-excludable in character and thus considered to be free of charge. This market failure to ensure the sustainable use of natural resources was identified in a publication by Garrett Hardin (1968) as 'the tragedy of the commons' and represents one of the major justifications for environmental policy in general and for PES in particular (TEEB 2009; Kinzig et al. 2011).

The concept of PES aims to address market failure by encouraging affluent beneficiaries of environmental services and owners of ecologically valuable resources to participate in a voluntary agreement that is based on a *quid pro quo* bargain (payment in return for the adoption of practices that ensure the sustainable provision of environmental services). The overall objective of PES is to make use of microeconomic incentives for sustainable land use management to address meso- and macroeconomic environmental and societal challenges with particular emphasis on global climate change and food security. The effectiveness of PES therefore depends on the way it interacts with the existing macro- and meso-economic institutions and the rights to use and trade land (van Noordwijk et al. 2012).

The PES concept has its theoretical roots in the early theory of neoclassical welfare economics of the 1970s (Pattanayak et al. 2010; Kosoy and Corbera 2010). It has been widely applied in different forms in developed and developing countries to address numerous types of environmental concerns. Yet, over the past decade, a gap has been observed between the theory of PES and its practical application in the field. This gap is due, on the one hand, to theoretical baseline assumptions that may no longer be appropriate for the global knowledge-driven economy of the twenty-first century and, on the other hand, to the particular political agendas of state and non-state actors that sponsor PES schemes but at the same time pursue other objectives.[3]

2.1.3 *Inconsistencies in PES Theory*

The literature on PES has grown exponentially in recent years and so has funding for PES projects all over the world (Fisher et al. 2009; Gómez-Baggethun et al. 2010). At the same time, there is widespread scepticism among PES scholars

[3] Often, the purpose of PES goes far beyond the effective and sustainable management of ecosystem services to include many other objectives such as poverty reduction, improved food security, preservation of cultural landscapes, ensuring decentralized settlement and increasing the quality of life of farmers in remote areas.

that point out the gap between PES theory and PES practice. Their criticisms of neoclassical PES theory are that it:

a) ...does not adequately address many well-known challenges in institutional economics and political economy, such as the problem of asymmetric information and the inadequate alignment of private incentives with the overall public interest (Kinzig et al. 2011; Kosoy and Corbera 2010; Pirard 2012a; Muradian et al. 2010; Pirard et al. 2010; Ferraro 2008; Aerni 2006, among others);
b) ...is based on a fictitious market because
   (1) demand and supply have to be organized and incentivized by an external mediator (Van Hecken and Bastiaensen 2010; Vatn 2010). The result is that farmers tend to adopt sustainable agricultural practices not because of the payment itself but because of many other forms of in-kind payments offered by the mediator. This helps farmers to increase not just the sustainability but also the productivity of their agricultural practices (FAO 2007). Yet, the heavy dependence on the external mediator also makes it unlikely that the system will function without him or her. There is therefore a high probability that the sustainability improvements will be reversed once external funding and support ceases.
   (2) the assumption that ecosystem services can be treated like a commodity that can be bought, sold and traded is built on shaky ground considering the complexity of the underlying social, political and biophysical relationships between humans and the environment (Kremen 2005; Norgaard 2010; Swift et al. 2004). A possible exception is carbon sequestration because carbon stocks are relatively easy to quantify, as they scale with the size of the area. Yet the transaction costs involved in complying with the strict standard criteria are widely considered to be unrealistically high for a market to arise from such a service in agriculture (Corbera 2012; van Noordwijk et al. 2012).
c) ...it tends to exclude the most vulnerable population members in rural areas who are confronted with uncertified tenure, lack of land rights, high transaction costs and high upfront investments (Lele et al. 2010). Certain PES schemes therefore focus on integrating the poor living in the most marginal and degraded lands (Pagiola 2007; FAO 2007). Even though this is a laudable aim, it is questionable whether the concept of PES is the right tool for addressing the combined development and environmental challenges of the poor since PES was originally designed for landowning farmers in developed countries. It would also represent a further step away from the 'polluter pays' principle which underpins the theory of PES. In the PES context, this principle has already been transformed into a 'provider gets/beneficiary pays' principle in the case of the RPE in agriculture. A pro-poor PES approach would represent a further shift from polluter pays to a 'pay the polluter' principle, because semi-subsistence farmers in Africa that are meant to be the providers of environmental services are often part of the growing environmental problems (deforestation, cultivation in riparian areas, soil impoverishment and erosion owing to a lack of nutrient replenishment) (Hanley et al. 1998). This perverse outcome may have been largely ignored by affluent

donors from developed countries because their PES approach is based on the implicit assumption that the driver of the depletion of natural resources and the destruction of environmental services is economic and technological change rather than poverty (MEA 2005). It has been shown, however, that poverty (combined with high population growth) can be as much of an enemy to the maintenance of ecosystem services as affluence (Beckerman 2002; Boserup 1976). This issue will be further illustrated in Chap. 3 by means of specific case studies in Kenya.

d) ... is based on a utilitarian approach that may undermine socio-ecological resilience and neglect other forms of valuation of nature and ethically informed nature conservation (Corbera 2012; McAfee 1999).

e) ...monetary incentives may crowd out existing pro-social behaviour in communities (van Noordwijk et al. 2012). This would again call into question the utilitarian framing of ecological concerns and highlights the importance of social capital as the crucial factor that drives pro-social motivations rooted in long-standing traditions and norms that favour collective action (Dietz et al. 2003; Cleaver 2002).

2.1.4 How to Make PES Ready for the Twenty-First Century Challenges

Do all these shortcomings mean that PES is failing to deliver? No, because there are many examples where PES seems to work well (as already illustrated in Chap. 2) and where the external sponsor has been able to exit without any indication that the stakeholders involved are abandoning the PES system created during the funding phase (the example of Costa Rica illustrates this well) (Pagiola 2008; Legrand et al. 2013). Yet, a closer look at these examples reveals that PES works for reasons that were not anticipated by PES theory and have little to do with the incentive provided by the payment itself. Instead, it may be the complementation of a voluntary PES scheme with some government initiatives that embed PES into an overall public sector endeavour to facilitate sustainable change (Chomitz et al. 1999; Wen et al. 2012; Legrand et al. 2013). The many forms of in-kind payments provided by the mediator of PES, as well as the social network created in the course of the implementation of PES projects, help to create new revenue generation opportunities for local participants (Namirembe et al. 2013; Prager et al. 2012). Finally, increased access to knowledge, capital and user-friendly technologies as a result of the presence of an external actor and the collaboration with previously unknown actors in the region may generate unexpected economic opportunities for local entrepreneurs wishing to create markets for environmental goods (Cohen and Winn 2007; Karrer-Rueedi and Trueb 2011).

In sum, the real success of PES may be rooted in its capacity to serve as a vehicle to create local sustainable business opportunities and thus also to stimulate sustainable economic change. Even though the role of local entrepreneurship and innovation has been acknowledged in the landscape approach to PES (Prager et al. 2012), the literature on its potential to make PES sustainable is still sparse. This stands in strong contrast to the literature on improving the effectiveness of conditional cash transfers (CCTs) that focuses increasingly on the need to make the achievements in poverty reduction and enhanced education enduring by investing more in the creation of knowledge-intensive off-farm employment (Rawlings and Rubio 2005).

Based on all these lessons learned, it makes sense to revisit the theory of PES from a theoretical and a practical point of view. There may be a need to move away from the comparative-static framework of neoclassical economics and embrace the insights of New Growth Theory. This theory acknowledges the positive welfare effects of the new products and services on society and the environment and it better reflects the lessons learned from past PES schemes in Africa as Chap. 3 will illustrate.

2.2 The Evolution of PES Theory

PES addressing sustainable management of the non-provisioning part of ecosystem services is a prominent incentive-based instrument in environmental policy and management, and is assumed to be in line with the legal principles of environmental law as well as with the neoclassical principles of environmental economics. Consequently, a reference to PES is found in almost all environmental treaties, laws, policy programmes and reports published by national and international environmental institutions (TEEB 2009; FAO 2007; AFD 2012).

PES is designed to provide incentives to landowners to use their natural resources in a more sustainable way, either by encouraging the production of positive externalities (adopting sustainable agricultural practices that reduce environmental degradation, water pollution and the loss of biodiversity) or the avoidance of negative externalities (e.g., to leave forest standing for the sake of the provision of environmental services such as water and carbon sequestration). The distinction is also referred to as land sharing (integration) versus land sparing (segregation) (van Noordwijk et al. 2012), or asset-building versus use-restricting (Wunder 2005). The former approach (land sharing–asset-building) is largely associated with European agri-environmental policy and the latter (land sparing–use-restricting) with US agri-environmental policy (Baylis et al. 2005) as also discussed in Chap. 1.

In neoclassical welfare economics it is argued that this type of internalization of externalities in economic decision-making is best done by giving a value to natural resources and thus eventually creating a market for environmental services in which the price reflects the actual demand from potential buyers and the available supply of the particular service from potential sellers (Arrow et al. 2004; Stavins 2005). As with any other monetary market, the PES scheme must define the service that is provided, create a standard unit of exchange for the particular service and identify buyers and sellers (see Fig. 2.1). This process allows the services provided to be perceived as objects that have the characteristics of commodities that can be

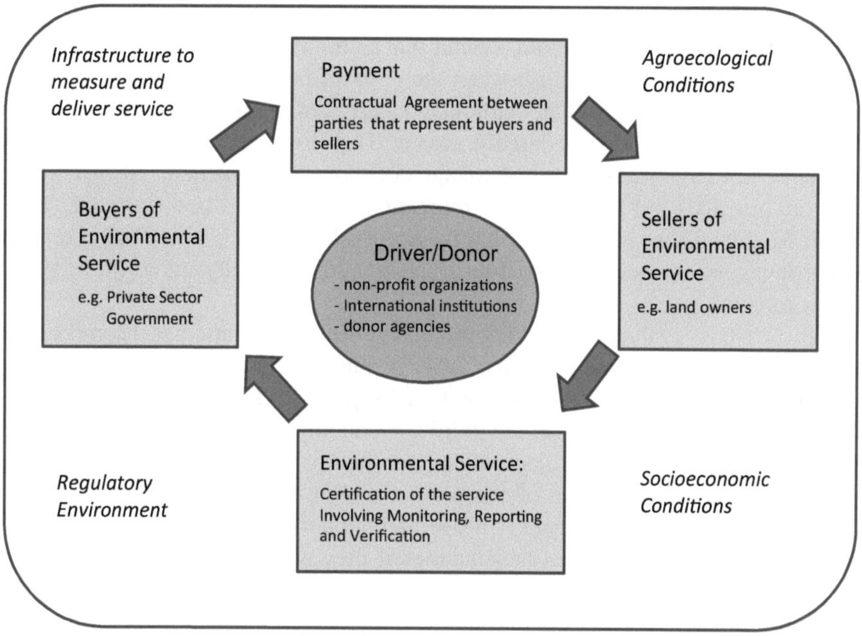

Fig. 2.1 PES as a market for environmental services operating in a particular environment

exchanged and that are subject to the laws of marginal utility (Pattanayak et al. 2010). This process of commodification involves three necessary stages:

a) reducing a particular ecological function to a well-defined and measurable ecosystem service;
b) defining a single exchange-value for this service;
c) identifying 'providers' (sellers) and 'consumers' (buyers) of these services in market or market-like exchanges.

2.2.1 *From Fictitious Markets to Evidence-Based Incrementalism*

PES hardly ever represents a real market that reflects effective supply and demand among market participants, since the main driver of a PES scheme is almost always either the government (as is the case in most developed countries) or an external donor (in developing countries).

The main objective of the mediator in the PES scheme is to ensure—with the additional funding and services provided—that the scheme is sufficiently attractive to get local buyers and sellers involved and keep them on board. It is also expected that the mediator will cover the transaction costs and create the necessary infrastructure to make the service measurable, transferable and replicable (MRV costs). Many PES scholars would counter that external funding is only necessary in the initial stage of a PES project when the costs of land-use change for sellers of environmental services may be particularly high and the infrastructure necessary for such a market to evolve largely absent. Therefore external funding may no longer be needed once the costs of change decrease and the benefits of sustainable farming practices and improved environmental services start to materialize (FAO 2007: 51–53). There is, however, hardly any evidence that classic PES is able to reach levels of financial sustainability that would eventually allow the external donor to exit.[4]

Most ex-post evaluations of PES projects reflect the situation at the end of the funding phase. But it remains unclear whether buyers and sellers continue with the conditional payments once the initial external support for the scheme ceases. In this context, one must also consider the political economy dimension of PES. External mediators that initiate PES projects often have their own political agendas reflecting the perceptions and interests of the particular constituency they represent back home. In other words, they must please their donors and supporters back home rather than the supposed local beneficiaries if they want to remain in the funding market for PES (Aerni 2006).

Moreover PES schemes operate in a complex environment with specific agro-ecological and socioeconomic conditions and where a lot of unknown factors may later turn out to be decisive in determining the fate of a PES project. Existing environmental regulation in a region where a PES scheme is being implemented may also either increase or decrease the incentives to participate in and continue with PES schemes (see Fig. 2.1) depending on the enforceability of the different policy measures on the macro-, meso- and micro-policy levels.

There is general agreement that PES needs to be evidence-based and many recent publications comparing the impact of different PES instruments in the field have provided interesting insights (Chen et al. 2014). Most scholars who use the evidence-based approach may contribute to incremental but temporary improvements of environmental management (Engel et al. 2008; Ortega et al. 2013) but they tend to be reluctant to question the baseline assumptions of PES theory that are not really concerned with the challenge of the financial sustainability once external support for such projects ends.

[4] A recent online discussion on the Global Forum on Food Security run by FAO revealed that there is growing discontent among PES practitioners because they feel that the classic PES schemes are not working and that hardly any of the failures have ever been properly evaluated and documented: http://www.fao.org/fsnforum/forum/discussions/pes

2.2.2 *Problems with Measurement and Asymmetric Information*

PES is considered to be a valid alternative to previous indirect conservation approaches (integrated conservation and development projects) that were found to have only a weak effect on environmental conservation. PES represents a direct conservation approach (based on direct payments) that was expected to be institutionally simpler, more cost-effective in the delivery of benefits, more effective in improving livelihoods of suppliers and more likely to provide new sources of finance for conservation (Ferraro and Simpson 2002). Consequently, PES concepts were tested in the field in many different parts of developing countries. Many scholars took the case of Costa Rica as a successful example of how to implement PES (FAO 2007; Pagiola 2008). PES policies were officially initiated there with the forest legislation dating back to 1996 (FONAFIFO, CONAFOR and Ecuador Ministry of Environment 2012). Landowners in Costa Rica received direct payments in return for adopting sustainable forest management techniques that do not affect the forest cover and provide valuable hydrological and biodiversity services. Even though the Government of Costa Rica acted as the main buyer, the PES scheme also relied on international loans and donations as well as international sales to finance the provision of environmental services (Chomitz et al. 1999). Yet, scholars who have reassessed the Costa Rica case recently have concluded that the positive trends in increased forest cover had already started to appear before the forest legislation was passed in 1996 (Miranda et al. 2003; Pfaff et al. 2008; Sills et al. 2008). This would question the 'additionality' criterion that is used as one of the major justifications for PES (Wunder 2005). 'Additionality' basically would require a PES scheme to document that the observed improvements of environmental services would not have taken place without the PES intervention. In other words, to be additional, pollutant reductions or land-use changes must occur in direct response to the payment. There are many other examples that are considered to be PES-like, such as biodiversity offsets, wetland mitigation or conservation programmes where the likelihood of additionality (vis á vis a preset baseline) is very low because conservation or ecological restoration in one site may merely be done to 'offset' land conversion elsewhere (Bennett 2010).

Criticism of PES has also been expressed with regard to the proxies used to measure an environmental service (forest cover does not automatically translate into improved environmental services because ecosystem function does not equal ecosystem service) (van Noordwijk et al. 2012; Tran et al. 2010; Kosoy and Corbera 2010). There is also the problem of asymmetric information in PES schemes (Ferraro 2008), which is also linked to the Principal–Agent Problem (PAP) (Akerlof 1970). The PAP argues that the agent (e.g., seller, contractor or manager) is better informed about a specific product, task or project than the principal (e.g., buyer, client, customer or shareholder). The agent uses this information advantage to further his or her own interests at the expense of the interests of the principal. In a PES project the PAP can be portrayed as an asymmetric information problem between the external donor or mediator of the PES project (principal) and the provider of the environmental service (agent). Since it is hard in almost all types of PES projects to exactly measure the improvement in the environmental service, there is an information gap that gives considerable space to agents (landowning sellers of PES) to make use of their information advantage. They will feel tempted to pass on to the mediator only the information that documents sustainable land use practices and to withhold the information that would call into question additionality (the agent would have done it anyway because of the law), economic efficiency (the agent would have known more cost-effective ways to address the environmental problem) or indicate leakage (the agent has moved polluting activities elsewhere). The PAP problem also exists between the mediator and its funding agencies, as well as between the mediator and the local buyers of the environmental service. It may thus be a double PAP in which the mediator also has an interest in highlighting the successes (e.g., trying to convince funders and buyers that the proxies chosen as indicators of the improvement of the environmental service do adequately reflect the actual situation) and concealing the problems of the project (e.g., high transaction costs, improvements that are offset by additional environmental pressure through population growth or lack of migration) (Aerni 2006).

The PAP is also linked to the problem of missing markets that result in shadow price effects in response to the introduction of a PES system, producing a rebound effect (Muller and Albers 2004; Sills et al. 2008). For example, in a subsistence economy, such as is often encountered in marginal rural areas in developing countries, choices depend on household-specific shadow prices, not market prices. As a result, a typical household may initially reduce the amount of labour dedicated to logging and other forms of resource degradation (fuelwood collection) in return for the payment for conserving the resources (direct effect). But the income from the payment is then likely to lead to an increase in the overall consumption of natural resources (shadow price effect).

2.2.3 *The Problem with High Transaction Costs and Unclear Property Rights*

The shortcomings in the design of PES described in the previous section, combined with other objections with respect to the narrow definition of PES, have led to an attempt to clarify when a PES scheme would be applicable and could effectively deliver results if certain criteria are met (Wunder et al. 2008). The criteria comprise 'enrolment' (attracting sufficient providers of environmental services), 'conditionality' (monitoring compliance and sanctioning non-compliance), 'additionality' (proving that sustainable land-use change would not have occurred without PES and that it is permanent), and 'quality of service provision' (ensuring appropriate quality and location).

However, these clarifications do not call into question the theoretical baseline assumptions of PES, which are based primarily on Coase's theory of the problem of social costs (Coase 1960)[5] and the associated principles of neoclassical economics.

If PES is defined according to Coase as a voluntary market-like transaction, low transaction costs and clear property rights would be necessary to make voluntary action the most cost-effective approach. But a spontaneous market-like and voluntary transaction between a buyer and a seller of an environmental service, as is assumed to take place within the narrow definition of PES (Wunder 2005; Engel et al. 2008) is unlikely ever to happen because the driving force of PES projects is almost always external and linked to a particular political agenda. Moreover, the potential buyers are almost never pure buyers[6] and the potential sellers are almost never pure sellers of an environmental service.[7]

2.2.4 *Moving Away from the Commodity-Fixation*

Proponents of the ecological economics perspective of PES have taken note of these theoretical weaknesses and embraced a broader definition of PES (Farley and Costanza 2010; Muradian et al. 2010; Vatn 2010). Muradian et al. (2010) argue that PES schemes must not necessarily be driven by economic incentives, that they do not have to be based on direct transfers but can rely on a mediator, and that the service itself does not have to resemble a commodity that is clearly quantifiable and identical in quality. Instead, they would redefine PES as a 'transfer of resources between social actors, which aims to create incentives to align individual and/or collective land use decisions with the social interest in a sustainable management of natural resources' (Muradian et al. 2010: 1205).

[5] Coase argued that if we lived in a world without transaction costs and clearly defined property rights, people would bargain with one another to produce the most efficient distribution of resources, regardless of the initial allocation. In such a situation externalities will be internalized in the form of side payments that reflect the cost of the perceived externality. Rising transaction costs and uncertain property rights will make it more difficult to internalize the externality in a voluntary agreement between the producer of a negative externality (e.g., polluter) and those affected by it. The consequence would be to resort to the law to discourage the polluter from offloading the social costs of his economic activities on the public at large (e.g., polluter pays principle).

[6] E.g. a company downstream that depends on clean water and is forced by regulation to limit the amount of untreated waste water produced may invest in a private wetland or a waste water treatment plant; it thus also becomes a provider of an environmental service.

[7] E.g. a farmer upstream may improve the sustainability of his or her agricultural practices and thus provide a particular environmental service to people that depend on clean water further downstream. But this does not change the fact that the farmer is first of all a user of environmental services (and thus should also be regarded as a potential buyer in consideration of the polluter pays principle).

This definition, however, includes a set of policy tools that have previously not been associated with PES, but represent market-based policy instruments (MBIs) in a broad sense. MBIs use markets in various ways, ranging from environmental regulations or private standards that change relative prices according to the environmental impacts (externalities) of production processes or land uses, to the creation of specific markets that limit the use of a given natural resource by organizing a trade in permits for the particular type of resource use. This helps explain why PES is also increasingly associated with all kinds of policy instruments and environmental initiatives, such as reverse auctions, voluntary and regulatory price signals, indirect incentives through large donor projects (e.g., water funds and livelihood projects, ecotourism, eco-certification schemes), subsidies for agri-environmental measures (including the RPE), compensation schemes for the provision of environmental services that are to some extent involuntary (e.g., see the case of China in Sect. 1.3.4), and others. Depending on the context, PES could therefore apply to most MBIs (Pirard 2012a). The results are often hybrid systems designed to improve the provision of environmental services while also creating economic opportunities for the local people. But what has so far been ignored is that you need not only policy instruments but also local entrepreneurs who make use of new opportunities to create new markets for environmental goods.

2.3 Hybridized PES Schemes to Reduce Poverty and Promote Development

A broad definition of PES may dilute some of the important requirements that must be met in order to count as PES but it also allows for more flexibility in efforts to adjust PES to local needs and circumstances. The resulting hybridized PES schemes often lead to unexpected new ways of doing things thanks to local initiative. In other words, PES schemes bring together people that usually do not interact (e.g., upstream farmers and downstream industrialists). The exchange does not just consist of a payment in return for a service but may induce the actors to find common interests in the creation of a local market for environmental goods. This could also result in initiatives related to local certification schemes, direct marketing opportunities for sustainable food producers, farmers becoming involved in tree nursing and selling, entrepreneurs making use of waste to produce new products, such as biogas, and others. It is a fact that a lot of environmentally valuable goods and services that are usually imported into the region can be produced by the locals. In such a case, PES becomes a vehicle for the creation of new markets and thus increases the likelihood that the PES project will be financially sustainable for reasons that were not anticipated when the scheme was being designed. Evidence in practice clearly suggests that this is the current trend, especially in Africa, but since research on PES is hardly ever linked to research on promoting entrepreneurship and innovation—and vice versa—scholars have a hard time understanding what is going on, apart from admitting that the theory does not correspond to practical experience.

2.3.1 *From Market Imperfection to Market Creation*

Market imperfections that lead to the unsustainable use of natural resources can provide market opportunities to address these imperfections effectively, provided that the public sector assumes the role of a facilitator by providing incentives for entrepreneurial innovations (Romer 1994). Substituting current unsustainable practices with technologies and supply-chain services that are able to minimize or nullify negative environmental externalities generate opportunities for new ventures and may eventually become a source of positive environmental externalities (e.g., remediation of polluted ecosystems and other ecological restoration technologies enable regenerative ecological capacity, which may result in increased species diversity and greater resilience) (Cohen and Winn 2007).

The advantage of promoting sustainable entrepreneurial ventures is that they may not just provide profits and create jobs, but also result in innovations that lead to more sustainable ways of living, and displace current, unsustainable means of production (Hart and Christensen 2002). As a result innovation helps to limit the trade-off between economic development and sustainable environment management. Thus, if a PES scheme leads to innovation, it may not only improve the environmental quality but also contribute to the reduction of poverty in the region.

2.3.2 *The Current Academic Debate on the Environment-Poverty Nexus of PES*

The existing literature on the environment-poverty nexus of PES has not looked at it from the perspective related to the creation of innovative new markets. Instead, the main argument has so far been that PES projects in developing countries should not just target the provision of a particular environmental service but should also be intended to improve rural livelihoods and agricultural productivity. Yet, in order to become eligible for a PES scheme that would jumpstart this development, farmers must meet certain criteria. Pagiola et al. (2005) argued that the poor would benefit from participating in such PES schemes if (a) they are located in an ecologically valuable area (thus making them eligible), (b) if initial payments exceed the provision costs (thus making them disposed to join the scheme) and (c) if they have secure property rights that would enable them to participate in bargaining. Once all these criteria have been met, small-scale farmers in degraded but ecologically valuable areas would become eligible and could thus benefit from the PES payment, which would then also reduce poverty, especially if it is not a use-restricting but an asset-enhancing PES, which adds in-kind payments to the direct payment for the environmental service. For example, an asset-enhancing PES scheme might enable PES providers to benefit from various kinds of extension services and planting material provided by the mediator. These in-kind payments enable farmers to increase agricultural productivity (FAO 2007). The question remains, however, whether PES is the right tool and whether it is sufficiently beneficial to ensure a permanent improvement of livelihoods in marginal areas. This is particularly relevant if the PES scheme does not result in any additional employment opportunities that are greatly needed in areas where farms tend to become smaller rather than bigger as a result of high fertility rates and limited economic opportunities outside semi-subsistence agriculture (see discussion of the Kenya Agricultural Carbon Project (KACP) in Sect. 3.2).

Moreover, the economist Tinbergen warned as early as the 1950s (Tinbergen 1952) that it is difficult to attain more than one objective with any single policy tool. His analysis would call into question the ability of PES to achieve both environmental quality and equity objectives because no strong correlation exists between the two (Zilberman et al. 2008). Even though that does not mean that the poor are unable to directly or indirectly benefit from PES schemes, the benefits are relatively small (Wunder 2008; FAO 2007). PES schemes can, however, be effective and even contribute to poverty reduction, if they are linked to public sector initiatives that add sticks to the carrots of PES (see Sect. 2.3.3), if they learn from the experience of existing conditional cash transfer programmes (see Sect. 2.3.4) and if they combine the RPE resulting from sustainable management practices with special rewards for local entrepreneurs who spot business opportunities in the effective and innovative delivery of environmental goods and services. The role of entrepreneurship and innovation in environmental management is increasingly being recognized in the landscape approach to PES and territorial development strategies.

2.3.2.1 Using Carrots and Sticks in PES

The pro-poor approach in PES watershed management projects is obtaining the widespread support of international donor agencies and international NGOs that seek to reconcile environmental goals with the overall objective of reducing poverty in mountain regions of developing countries. However, even though such projects may achieve environmental improvements and poverty reduction in the short and medium term—mainly through in-kind payments rather than the PES cash transfer (Leimona et al. 2009; FAO 2007), it is not certain whether the projects would be financially sustainable once the main donor withdraws, considering that they are the primary providers of such in-kind payments. Moreover, pro-poor PES projects that are primarily based on carrots for the poor (pro-poor rewards) to comply with sustainable agricultural practices upstream may be failing to reduce upstream pollution because they provide the wrong incentives. Upstream farmers may welcome an additional payment for environmental services as long as the external donor takes charge of the introduction of sustainable agricultural practices, but they may not feel accountable if the project fails to improve the environmental service in any significant manner. This has a lot to do with the reframing of the 'polluter pays' principle into the 'polluter gains' principle (Vatn 2010) or the 'pay the polluter' principle (Hanley et al. 1998). Many affluent donors from developed countries nevertheless continue to rely on carrots alone because they consider that PES in developing countries is also a tool to reduce poverty. Moreover, they start from the assumption, prevailing in most affluent countries, that technological and economic change (taking place in industrialized downstream regions) are the main drivers of the depletion of natural resources and the destruction of environmental services (MEA 2005).

Yet, evidence from developing countries suggests that poverty (combined with high population growth) may be the main driver of water pollution, deforestation and environmental degradation (Boserup 1976; Beckerman 2002; Hollander 2003; Gilbert 2012). A tacit change from the 'polluter pays' to the 'polluter gains' principle resulting from the desire to reconcile poverty reduction and environmental goals may therefore benefit neither the poor nor the environment in the long term. Emerging economies that feel under more pressure to tackle the challenges effectively tend to be more reluctant to try to achieve two goals with one instrument by sacrificing the fundamental principle that provides the basic rationale for applying the instrument. In China, for example, the PES approach is dealing increasingly with transboundary problems involving several political districts. This makes it even more difficult to implement a classical PES scheme because it may conflict with the different development plans and policy strategies in the respective districts. As a consequence the Chinese government has started to complement the carrots for environmental services providers (large grants from the central government to improve environmental services with the option of additional cash transfers from the downstream region) with corresponding sticks in the case of non-compliance (Wen et al. 2012). The agreement in this government-led PES scheme involving a downstream and an upstream region is that water quality must improve beyond the baseline level after a certain period of time if the downstream users are to pay a preset amount to upstream providers. Conversely, if water quality has deteriorated over time, upstream providers must compensate downstream users correspondingly. Such a scheme could become an interesting option for other regions in developing countries where the classic situation of an urbanized and industrialized downstream region and a rural and poor upstream region does not necessarily apply.

In cases where the upstream providers of PES are simply too poor to apply a sticks and carrots approach, the reversal of the polluter pays principle may be justified provided that the public sector acts as a facilitator and enforcer of sustainable change. It then becomes a mandatory PES based on the enactment of regulation that builds upon a successful pilot phase of a voluntary PES. This has been successfully tried in Lombok where a voluntary PES project was initiated with the support of the World Wildlife Fund (WWF) in 2003 to prevent the degradation of water supplies through afforestation of the region around the Rinjani volcano in the northern part of the island (Pirard 2012b). The PES scheme eventually became mandatory when regulation was enacted to involve most of the resident water consumers as 'buyers' (companies and households mostly located in Mataram City). Seventy-five per cent of the fees collected are allocated to payments through PES contracts while the remaining 25 % are reserved for overhead costs in the district budget. A multistakeholder body (including WWF) has been established and authorized to collect the financial resources and enforce the contracts with the providers of the environmental service. Even though the PES-like contracts with providers have not yet been finalized, the Lombok case serves as an interesting example to illustrate how a high degree of convergence can be achieved between public policies and PES. Permission is required from the service beneficiaries or the authorities to log planted trees. At the same time, landowners and users can generate incomes from standing forests and agroforestry. Even though most forests in Indonesia are state property, the Hutan Kenasyarakatan programme, launched by the government in 2001, provides limited rights to resident populations under conditions of sustainable forest management. This allows residents to undertake productive activities as long as the conditions are respected. The programme is popular (185 ha in the Rinjani area have already been granted this status) and very compatible with PES since one of the conditions for obtaining the forest user rights is to rehabilitate land through tree planting elsewhere. Few applicants would have the opportunity to do so without PES becoming the financing vehicle (Pirard 2012b). Obtaining forest user rights in return for the rehabilitation of land through tree planting elsewhere is opposed by those who consider PES to be a use-restricting instrument to preserve ecosystems in a 'primordial condition'. This objection does not take into account, however, that many valuable species of fauna and flora have been proved to be able to live in secondary forests as well as primary forests (Shell and Meijaard 2010). Ecological restoration has therefore proved to be a valuable tool for 'asset-building' in PES and should be prioritized if one seeks large-scale implementation of PES (Pirard 2012b).

2.3.2.2 Combining PES with Conditional Cash Transfer

Another way in which PES may be made more effective is to take into account the lessons learned from CCT schemes in Latin America (Rawlings and Rubio 2005). CCT policy tools have proved to be an effective means for promoting human capital accumulation, child health and food security among poor households in rural areas as illustrated by Mexico's *PROCAMPO* and *Opportunidades* Programmes (Winters and Davis 2009). However, investment in education combined with limited options for knowledge-intensive off-farm employment tends to keep the newly educated trapped in low-productivity occupations (Levy 2007; FAO 2012)]. This suggests the need for complementing investment in education with investment in entrepreneurship and innovation, which would reduce the severity of rural poverty through the creation of employment opportunities (Winters and Chiodi 2010).

Despite being based on a similar concept of conditionality, most PES programmes have not been subject to evaluations of a similar scientific rigour to those of CCTs (Ferraro and Pattanayak 2006; Pattanayak et al. 2010). There is also a difference between investing in human capital (where CCT is applied) and investing in natural capital (where PES is applied). The positive impact of investing in human capital (sending children to school, ensuring adequate health care) made possible through CCT incentives enables a household to build up assets over time. In other words, the incentive to comply with CCT conditionality is high because the private interest of the participating household is well aligned with the public interest of having a more educated and less impoverished population. In the case of PES, the positive impact of the adoption of sustainable agricultural practices in return for the payment may be attractive if these practices also increase agricultural productivity and generate other sources of income (FAO 2007). However, the case here is much less straightforward since the recommended sustainable practices are often labour-intensive and dependent on outside assistance and materials in the form of in-kind payments provided by the meditator of the PES scheme. Moreover, the payment itself is too small to be invested in long-term asset-building. If PES encourages investment in labour-intensive sustainable agricultural practices at the expense of investment in children's education, it may also be in conflict with the need to generate additional household revenues from the educated children that find work outside farming or set up a new business. In other words, the public interest in improving environmental services might be out of line with the participating households' interests in building up assets over the long term. Moreover, measuring the compliance of PES practices and its overall relationship to the improvement of ecosystem services in the region in general is much less straightforward than it is for CCT.

CCT and PES could however be mutually reinforcing if PES were to come on top of CCT, meaning that once a household complies with the conditionality of CCT, it would also become eligible for PES as a source of additional income that would not be at the expense of additional household labour in the field. Yet, CCT as a social policy tool in conjunction with PES as an environmental policy tool would not guarantee a sustainable path out of poverty unless they are also attached to the building up of an entrepreneurial infrastructure. Such an infrastructure would be necessary to allow the better educated future generation to set up a business and thus create off-farm employment. Off-farm employment, in turn, could absorb the population surplus in agriculture, facilitate structural change, generate more revenues for rural households and improve rural–urban linkages. The awareness of this need to invest in the economic activities of the rural poor has been most successfully tackled with the programme *Oportunidades Rurales* in Colombia. This programme helps indigenous people, afro-Colombians and other disadvantaged groups in the countryside to become the rural entrepreneurs of tomorrow by providing access to technical assistance and business and financial services to enable them to develop profitable businesses that will also serve their communities. The primary objective is to finance innovations that will be relevant to rural people. Through the initiative, almost 4000 young people have received training in business administration, marketing and investment. To emphasize the importance of saving, the project also provides 50 per cent matching funds for every peso saved. Deposited in local banks, these savings are then available for lending to other entrepreneurs in the community. To gain access to funding for technical assistance, each competitor prepares a proposal and a budget, which are judged by experienced micro-entrepreneurs. The maximum loan per family is US$700, and microenterprises can receive up to US$40,000 (IFAD 2011). Such projects investing in entrepreneurship and innovation in the countryside can easily be linked to social and environmental goals by providing special financial incentives for local entrepreneurs to invest in goods and services that would have a high public-good character for the region if they were available (examples such as microinsurance for farmers in the Mount Kenya region, biogas production from slaughterhouses in Kiserian and wetland construction businesses in Thika and Naivasha are documented Chap. 3 on Kenyan field studies).

2.3.3 The Role of Entrepreneurship in the Landscape Approach to PES

The potential of entrepreneurship to make PES financially sustainable has been recognized in the context of the need to shift from a field-level approach in which each individual farmer is rewarded for conservation efforts, to landscape-scale coordination based on landscape designs (OECD 2001, 2011; Prager et al. 2012; McNeill et al. 2012). Landscape designs have two critical components: composition and configuration. Landscape composition refers to aggregate land cover or land use in a landscape whereas configuration refers to the spatial arrangement of land cover or use in a landscape (Turner et al. 2001). Land use change is the primary driver of both landscape configuration and composition and the provision of ecosystem services. The challenge is to create institutional incentives for service provision that extend across property boundaries. Current PES incentives reward individual farmers' behaviour but provide little, if any, inducement for cooperative conservation across property boundaries. Goldman et al. (2007) propose three incentive designs to cope with the cross-boundary challenge, namely:

a) The 'cooperation bonus' incentive scheme, which rewards conservation even in the absence of cooperation, but adds a bonus for cooperation. The purpose of the bonus is to encourage connectivity of conservation across the landscape in multiple ways. The incentive rewards particular landscape configurations that take into account contiguous acreage and thereby most effectively target local services such as pollination (Shogren et al. 2003). The incentive could also reward landscape composition by providing a bonus for particular ecologically valuable plants or trees. Since there is so much heterogeneity in crop type, opportunity cost of forgone acreage and landscape type, designing one generalized incentive would, however, be time-consuming if not impracticable.

b) The ‘entrepreneur’ incentive scheme, which rewards creativity by providing landowners with a chance to create their own landscape designs promoting ecosystem services and cooperation on any scale. The entrepreneur (a farmer or a third party) may come up a landscape that is both feasible and effective for ecosystem service provision. His task is then to cooperate with landowners on possible and feasible land configurations. The reward scheme would pay for a total package that a cooperative of landowners presented in a competition. Those applications with the best design, as ranked according the established criteria, would receive the most funding.
c) The ‘ecosystem service district’ scheme, which is only partially voluntary and forces all the stakeholders involved to cooperate. The creation of an ecosystem service district may be necessary to endow districts with sufficient financial support and authority to design and enforce the provision of specific ecosystem services through appropriate incentives. Ecosystem service districts would be useful in landscapes where the participants might be less familiar with or even mistrustful of one another. The district incentive provides a very formal, regulated means of interaction that complements the more voluntary PES schemes in many ways (Pirard 2012a).

The main objective of creating such incentives is to minimize the public-good problem and to create cross-boundary, landscape-scale conservation that is financially sustainable. Goldman et al. (2007) illustrate how entrepreneurship may play a central role in the landscape approach, especially if the entrepreneur is a trusted and integrated member of the relevant community; this might help diminish the transaction costs associated with lack of trust in externally funded programmes (Kraft et al. 1996).

2.3.4 PES as a Vehicle for the Creation of a Market for Environmental Goods

There are several examples in which a PES scheme, whether it was initiated through environmental law or through voluntary initiatives, has served as a vehicle to create a market for environmental goods that is also scalable and can be tailored to local circumstances. Concrete examples illustrated in the following paragraphs are (a) companies that are specialized in the construction and restoration of wetlands, (b) companies that focus on the delivery of a package of environmental goods and services that enable farmers to enhance biodiversity and water quality on their farms and (c) companies and not-for-profit organizations that create eco-labelling or ecotourism schemes attached to the conditionality of sustainable agricultural practices.

(a) Well-functioning wetland mitigation banks illustrate the type of creativity that such incentives could inspire (see Sect. 1.2.3). Wetland banks are established to offset the destruction of wetlands in other areas. Ideally, such a bank creates, expands, or restores a wetland to provide compensatory wetland acreage for developers that are required to buy 'credits' in the wetland bank. Again, one might question the additionality of wetland banks since they are mainly about compensating for the loss of wetlands elsewhere (see Sect. 1.2.3). But they have had the great benefit of creating a market for environmental goods resulting in a 'wetland construction business' that is not only active in developed countries that require compensation for lost wetlands by law but also ensures that export-oriented agribusinesses in developing countries comply with private standards by setting up their own private wetlands that minimize the release of harmful effluents into the surrounding environment (as illustrated by the case of the flower export business and its private standards in Sect. 3.7). An entrepreneurial incentive system may encourage a domestic wetland construction business that adjusts and tailors the technology to meet local needs.

(b) The Common Agricultural Policy (CAP) of the EU consists of two pillars that are currently under reform to make them more effective and financially sustainable: Pillar 1 primarily finances direct payments which will be increasingly linked to the conditionality of sustainable agricultural practices. Pillar 2 funds rural development and will have a focus on enhancing innovation and entrepreneurship in rural areas. Both pillars will create incentives to develop markets for environmental goods. One successful example was illustrated in Sect. 1.3.4. The agro-chemical company Syngenta, which had already invested a significant amount in research on sustainable agro-ecosystems in the 1990s, started to develop concrete ingredients (seed packages) that help farmers to manage sustainable agro-ecosystems attracting more bees and birds, increasing biodiversity and improving the water quality. The CAP reform was necessary to allow them to develop such a market for environmental goods and services. It became commercially viable to assist farmers in their efforts to meet agri-environmental cross-compliance criteria for which they received remuneration on the EU and the national level. Since the agro-ecological conditions differ from one region to another, Syngenta collaborates with local multipliers of valuable planting material, universities and farmer groups as well as NGOs, government bodies and food-chain partners. The additional biodiversity created on-farm as a result of the goods and services offered by Syngenta also helps to make crop cultivation more productive, minimize the impact of intensive agriculture on the environment and, at the same time, makes the land more attractive for agro-tourism.

(c) There are already many examples of instances where individuals and companies have spotted new business opportunities related to eco-labelling and other certification schemes to reduce the market frictions that prevent market participants from properly valuing ecosystem services (Lockie 2013). However, certification schemes can also make business sense for the big food companies in the food chain. These big retailers help farmers to ensure the provision of environmental services more cost-effectively by offering them access to new premium markets for their agricultural products, provided that they comply with the sustainable practices that assure the quality of environmental services (Gaupp 2013). The resulting knowledge transfer and the increased demand for assistance in the compliance with ecosystem services also creates opportunities to set up local businesses that have the potential to be scaled up, thus creating more off-farm employment and jumpstarting a process of social empowerment.

The trend towards markets for environmental goods also indicates that the holistic approach as advocated in many PES systems does not necessarily mean that every farmer has to do everything himself. Farmers who really prove to be innovative and efficient in ensuring that their agricultural practices lead to good harvests and do not impair the quality of environmental services should be able to offer their skills to other farmers in return for a payment, be it from the recipient farmers themselves or from the mediator of the PES scheme. If local farmers lack the purchasing power to buy such services, a special voucher programme could be designed that would enable them to access these services more easily.

2.4 From Practice Back to Theory

The theory of PES identifies the degradation of ecosystem services as occurring because such services are mostly provided by nature and are therefore considered as public goods that are non-excludable and non-rival. However, they are common pool resources in the sense that their value eventually decreases with increasing use. In other words, the depletion of common pool resources must be considered as a 'negative externality', understood as an unintended consequence of the primary economic activity. Garrett Hardin recognized the problem with common pool resources in his seminal article in *Science* (Hardin 1968) entitled 'The tragedy of the commons'. The impairment of the provision of environmental services resulting from the tragedy of the commons involves 'off-site' impacts (affecting society and the environment at large) and on-site impacts (affecting farmers directly) (Karlen et al. 1997).

It is generally assumed that the growing demand for food, fibre and fuel resulting from population growth, increasing affluence and greenhouse gas emissions in emerging economies will increase the pressure on the provision of environmental services (FAO 2011a).

2.4.1 *The Baseline Assumptions of Welfare Economics and Its Flaws*

In its review of the state of global ecosystems in 2005, the Millennium Ecosystems Assessment report (MEA 2005) concluded that progress in the increase of provisioning services (e.g., increasing food production) in the twentieth century thanks to industrial agriculture came at the expense of a decrease in the regulating and supporting services of ecosystems (e.g., water quality, biodiversity, climate change) that was negatively affecting the well-being of humankind and even endangering the delivery of provisioning services for future generations. These negative externalities of the agricultural economy for society and the environment must be addressed through the design of institutions that allow for the internalization of these negative externalities by assigning a value to intact ecosystem services.

Many welfare economists made exactly the same arguments during the Cold War period and designed the environmental policy instruments that are still largely applied today (Stavins 2005). Ronald Coase (1960) argued, for example, that socially suboptimal situations (e.g., decreasing quality of environmental services) can be addressed by assigning property rights to the use of natural resources and subsequently lowering the transaction costs for the trading and exchange of such rights. This was the beginning of the idea of converting ecosystem services into tradable commodities. This should eventually facilitate voluntary market-like transactions in which buyers (mostly from elsewhere) who are concerned about the regulating and supporting services in a valuable ecosystem pay the sellers (mostly local users of natural resources) to restrict their use of natural resources for the production of provisioning services in return for monetary or non-monetary compensation. The payments would then be conditional upon maintaining a sustainable ecosystem management that provides a well-defined environmental service. Even though assigning property rights provides important incentives to conserve supporting ecosystem services in order to ensure the provisioning services upon which economic activities rely, the exchange of these rights, and consequently their commodification, is more difficult, as already discussed in Sect. 2.1.3. The property rights-based approach is also attracting increasing criticism from human rights activists who argue that such schemes tend to impose the environmental and social land use preferences of affluent communities that have the means to acquire property rights, on poorer marginal communities that may be forced to sell such rights even though they depend on the provisioning services of their ecosystems and insist on their historical right to make use of them (McAfee and Shapiro 2010; Ankers 2001).

These economic principles may not be neutral from an ideological point of view (Gómez-Baggethun et al. 2010) and they tend to ignore the historical dimension that shaped the unquestioned baseline assumptions for the principles (Aerni 2009). The rise of the principles of neoclassical economics in public policy must be understood in the Cold War context (Gilman 2004). The many comparative-static elements in its analysis might have made sense in the bipolar geopolitical situation after World War II, but appear to have become increasingly inadequate to explain the dynamic processes of today's global knowledge economy, which is characterized by rapid catch-up growth in developing countries and increasingly short life-cycles of innovation, including in agriculture (Warsh 2006; Jones and Romer 2010; Magnier et al. 2010; Aerni 2011b). Even though attempts are being made to replace neoclassical economics with more evolutionary and ecological approaches to PES theory (Sabau 2010), the welfare effects produced through the introduction of new goods and services, highlighted in New Growth Theory (Romer 1990, 1994), are still not taken into account. As consequence, the contribution of innovation to the financial sustainability of PES schemes is not addressed.

The neoclassical welfare economics approach may be increasingly inadequate for explaining decision-making in contemporary politics and economics because there are several flaws associated with the underlying theoretical concept:

1. It is based on a very pessimistic view of the market economy, which is best reflected in the laws of diminishing marginal utility and returns. These laws assume that no new goods will come into being whereas the returns from the production of existing goods as well as the utility derived from their consumption will diminish over time. Such assumptions would imply that the market economy would eventually come to a standstill (Schumpeter 2008). Yet, in reality the opposite was observable in Europe during the twentieth century and is already being observed in Asia in the twenty-first century.
2. Its goal is to maximize welfare through the optimal allocation of scarce resources (Pareto criterion) and adequate compensation of the identified losers (Kaldor-Hicks criterion). These two criteria are used to implement the normative policy objectives set by the 'social welfare function', expressed in the form of aggregated social preferences. Yet, there is no such thing as a social welfare function because individual preferences and the individual utility functions on which they are based are socially constructed, unstable and highly diverse. Moreover, 'willingness to pay' surveys, which are used to measure preferences, have many flaws because they ignore most of the findings obtained in experimental research about affect heuristics (depending how you frame the question you will obtain the value you desire as an expression of social preference) (Lichtenstein and Slovic 2006). The claim to know the aggregate social preference, however, is very convenient for opportunistic political actors who try to gain the public's favour by claiming to act in the 'public interest' that is supposed to reflect aggregated normative preferences.
3. Its underlying concept of a rational social planner that is in charge of efficiently implementing the normative goals set by the social welfare function for the sake of the public good may be applicable to authoritarian regimes (benevolent dictator) but are hardly compatible with the diverging interests pursued in the political process of contemporary democracies (Buchanan and Tullock 1962). The concept was crucial to the centrally planned socialist economy and continues to be widely applied in agricultural, environmental and development economics. The failure of communism, however, suggests that such a naïve

view of government as a responsible and responsive social planner often results in large-scale mismanagement of public resources rather than social welfare (Hayek 1988).

4. Its implicit assumption that public goods must be exclusively managed by the public sector because they are non-rival (can be used by many without losing value) and non-excludable (cannot exclude individuals from their use), is flawed. First, almost all public goods created by governments are funded with tax revenues by the residents of the country that make use of the goods. Second, the private sector is also producing positive externalities for society that have public-good character (e.g., more choice, employment, technological innovation, wealth and the more efficient use of natural resources) and technological evolution makes it increasingly effective to let the private sector help manage public goods (Heal 1999). The concept ignores the fact that positive and negative externalities are general by-products of any human action no matter whether it is a public or private sector action (Buchanan and Tullock 1962).

These inadequacies in the assumptions of neoclassical welfare economics were identified and rebutted by many of the leading economists of the past five decades representing different types of economic analysis. Yet apparently their insights have only had a minimal impact on mainstream textbook economics as David Warsh notes in his book *Knowledge and the Wealth of Nations* (2006). Nevertheless, there are viable alternatives to neoclassical economics that have already been successfully adopted in the economic policies of many emerging economies. One such alternative is Paul Romer's New Growth Theory. Its emphasis on the role of institutions that facilitate sustainable change is able to re-cast government into the role of facilitator of change and to calculate the degree of public welfare effects generated through the introduction of new goods and services (Romer 1994)—something that welfare economics is unable to do. Charles Jones and Paul Romer have also reframed the stylized facts defined by Nicolas Kaldor in 1961 to explain economic patterns and the driving forces of economic growth. Whereas Kaldor focused particularly on the role of the input factors, labour and capital and the role of interest rates in investment in capital goods, Romer and Jones look at the role of ideas, institutions, population and human capital and how they drive economic growth through innovation (Jones and Romer 2010). Dani Rodrik, in turn, identified two driving forces for sustainable economic growth: the development and accumulation of fundamental capabilities and structural transformation through new platform technologies. He argues that the policies needed to accumulate fundamental capabilities and those required to foster structural change through emerging transformative industries naturally overlap, but are distinct. Public investments in a broad range of skills, education, administrative capacity, and governance can be combined with narrower, targeted remedies to foster structural transformation and at the same time address its challenges. The conclusion of his analysis is that successful, heterodox policies work precisely because they compensate for weakness in fundamentals (Rodrik 2013).

2.4.2 *Innovation as a Generator of Positive Externalities in PES*

Once some of the theoretical baseline assumptions have been abandoned, PES schemes become increasingly receptive to the positive externalities that innovation can generate for society and the environment. A market for environmental goods is not simply about a buyer and a seller of a natural good that would otherwise be depleted, but also about innovations that allow people to use the natural resource at stake more sustainably without the need to sacrifice the economic growth that is so vital, especially in poor countries with high population growth rates and few off-farm employment opportunities. Such innovations are based on non-rival knowledge, the only input that is not subject to the laws of scarcity (see Box 1).

BOX 1
Why innovation generates positive externalities

Knowledge is not subject to the laws of scarcity and diminishing returns and since it is a non-rival good its increased use does not undermine its value but actually enhances it (Romer 1990) If it is employed either to continuously improve existing goods and services or to create discrete innovation that result in entirely new goods and services, it can generate value not just for the company that applies the new knowledge but for society as a whole. This is especially so if innovation is spread through international trade. In other words, depending on the particular type of innovation, it may generate important positive externalities for society at large. The resulting welfare gains generated through the introduction of new goods and services were demonstrated by Paul Romer (1994), but remain largely ignored in neoclassical economics. As a consequence, the public-good character of private goods and services has hardly ever been discussed in the discipline of economics (apart from the literature that discusses the positive economic spillovers of innovation clusters). Technological change not only enhances the potential of the private sector to contribute to the creation of public goods, but also to minimizes public bads. Effective public policies make use of public–private partnerships to maximize the public benefits and minimize the public risks of private sector activities (Heal 2000), especially by providing incentives for companies to invest in new knowledge that helps to solve problems that are of public concern. Once new knowledge is produced in the form of ideas, instructions, protocols, or designs that lead to innovative products and innovative solutions, this knowledge assumes the character of a non-rival good. Consequently, complete property rights and perfect competition that work so well in a world consisting solely of rival goods may no longer be desirable because they are likely to stifle innovation and prevent the optimal allocation of resources. Yet the increasing returns resulting from the creation of product innovation will only be realized if a company can reasonably expect to be able to reimburse the fixed costs for research and development (R&D) by selling the product for a price that is above the marginal cost of production per unit. In other words, the company must be able to extract a temporary monopolist rent (profit) which enables it to remain competitive, not in a market of perfect but of monopolistic competition. It is this price-setting power in monopolistic competition that provides the main incentive to make use of new knowledge and invest in new goods and services that produce not only profits, but also welfare effects for the public at large. A single price, set by many buyers in a market of perfect competition, is unable to simultaneously allocate goods to their most efficient uses and provide the appropriate incentives for innovation.

Innovation does not have to occur locally but can be adopted from elsewhere and then merely needs to be tailored to local needs.[8] However, for that to happen there needs to be a local entrepreneur that can count on institutional and financial support to successfully commercialize this type of sustainable innovation and thus create a market for environmental goods that generates increasing returns not only for the entrepreneur, but also for the local community and environment at large. In other words, a green economy requires green innovation that is successfully

[8] In this context, the term ‘innovation’ should not be understood as a mere idea or invention but as the process by which such an idea or invention is translated into a good or service for which people are willing to pay.

commercialized by green entrepreneurs (Farinelli et al. 2011). In this context, governments need to create economic incentives that help such green entrepreneurs to reduce risk and uncertainty when deciding whether to invest in the development and commercialization of a new good or service that is likely to generate positive externalities for the region and therefore also has public-good character. Such incentives may also be directed towards other stakeholders in academia, government and civil society to encourage them to collaborate with the entrepreneur in his or her effort to successfully launch the product innovation, as is already happening in the agricultural economy of New Zealand (Aerni 2009). Incentives may consist of financial rewards for the contribution to the generation of the positive externalities that result from the particular product innovation.

2.4.3 Governments as Facilitators of Public–Private Partnerships

Even if there exists effective public–private partnership to develop and disseminate a product innovation with a public-good character, the purchasing power of the expected beneficiaries might still be too low to generate sufficient revenues for the consortium to cover the fixed costs invested in the innovation. In this case governments may also have to help to create the necessary demand through advanced purchasing agreements (Berndt et al. 2005). They can also create a market by offering vouchers to poor households, which increase choice and provide access to a preferred product innovation (Blomquist 2004). Such policies may not be associated with PES but nevertheless contribute to the generation of positive externalities that help to reconcile environment and development objectives through innovation. Rather than expecting that PES can indeed function as a self-sustaining voluntary agreement between a buyer and a seller of an environmental service, as orthodox neoclassical economics would suggest, PES may instead serve as an initial vehicle to facilitate such markets (generating positive externalities). This is because PES schemes may contribute to the building of an enhanced social network that also includes investors from urban areas and universities. In turn, such a network would help to increase access to user-friendly and affordable alternative technologies (e.g., solar lamps, biogas stoves), increased agricultural productivity through improved seeds and management techniques, premium prices or input discounts for sustainable agricultural practices, and new mobile phone applications that help to lower the transaction costs of contract enforcement and enhance access to relevant market information (Kodhandaraman and Daniel 2010; Karrer-Rueedi and Trueb 2011).

2.4.4 Moving Towards Progressive Agricultural and Environmental Policies

Today's environmental and agricultural policy instruments are mainly designed to regulate or prevent unsustainable change, but there are few policy tools that also help to facilitate sustainable change. An implicit baseline assumption in many RPE/PES schemes is that technological and economic change in industrial agriculture produces negative externalities, while the positive externalities that are generated by traditional agricultural systems, especially if they are managed in a sustainable way would be ignored. Consequently, industrial agriculture is expected to pay for the negative externalities it generates, whereas traditional agricultural systems must be remunerated for the positive externalities they create by adopting sustainable agricultural practices. Agribusiness companies downstream are therefore often identified as buyers of watershed services while the small-scale farmers in the upstream region are considered to be the potential sellers of such a service. But does this view still hold? Agribusiness might produce innovations, markets and employment opportunities that could also generate positive externalities for small-scale farmers upstream. And what if the payments have the unintended side effect of slowing down structural change in agriculture in poor regions that desperately need it? If traditional agricultural systems selectively embrace technological and economic change in Africa (if mechanisms to facilitate access and capacity development are in place) the resulting productivity increases may help African countries to better cope with high population growth rates, especially in poor rural areas that are characterized by extremely small average farm sizes. Productivity increases result in more rural revenues and more investment in off-farm employment which again allows for growing farm sizes. In return, if farmers remain stuck in traditional systems, farm sizes may become even smaller due to lack of structural change. The result is decreasing productivity and lack of off-farm employment opportunities in the private sector, and thus increasing food insecurity and environmental degradation. No matter whether or not sustainable practices have been adopted on-farm, if the farm size shrinks to a level at which it can no longer generate any surplus, then there will be no more means available to replenish the soil or buy improved seeds. In other words, poverty can be a bigger threat to sustainability than agricultural modernization.

2.4.5 Making Entrepreneurship and Innovation Incompatible with Local Traditions

Many organizations that are engaged in nature conservation in developing countries may not perceive PES as a market. They believe that nature has an intrinsic value that cannot be properly reflected in a market price (Van Hecken and Bastiaensen 2010). However, since it is the external funding that largely supports the conservation of environmental goods that are believed to have a large non-market value, the maintenance of the intrinsic value of nature must still be regarded as a product of the global monetary economy. Moreover, the fact that money is spent on the provision of particular goods or services with positive externalities automatically signals to local people that there are opportunities. Consequently, it influences people's day-to-day decisions in expectation of particular benefits. Spending money on something implies in the classical sense that there is a seller and a buyer of this 'something'; and because both parties seem to accept money as a means of exchange it fulfils, whether intentionally or unintentionally, the criteria that define a market—even if the purpose of a project is to support traditional barter-trade in rural villages (which would otherwise be abandoned).

This touches upon the intrinsic paradox of western food sovereignty and indigenous people's rights advocates who are essentially reflecting the values of an affluent post-materialist society (Aerni 2011a). They imply that their values are shared by indigenous tribes whose members must feel equally disillusioned with the global economic system that is trying to undermine their cultural traditions and get hold of their natural resources. By default, they also assume that all the traditional institutions of indigenous communities are sustainable and that any attempt to integrate indigenous people into formal economic activities may result in subjugation and deprivation. PES and other forms of RPE are thus often opposed for their utilitarian approach and the requirement for formal institutions that protect private ownership. It is argued that PES schemes would ignore the spiritual dimension of a landscape and its cultural heritage. The belief that indigenous tribes clearly distinguish between the utilitarian and the spiritual dimension of natural resources, however, is another Western projection. Moreover, the static view of indigenous culture is misleading because most of the traditional practices of indigenous tribes have their roots in the exchange with other communities (Diamond 2012). Keeping people and their practices isolated from outside influence may therefore do a disservice to indigenous tribes, especially if population growth (resulting from previous aid and trade activities) makes it difficult for traditional institutions to cope with the new economic, social and environmental challenges. A community may completely decouple itself from trade and exchange with the outside world and thus be perfectly autonomous in its right to control, produce, and consume local food. But does that really mean that such a community is sustainable? What about the next generation that may not necessarily want to be mere custodians of cultural heritage but might prefer to combine and expand it with new practices and insights from elsewhere? If the community members decide to invest in a business and thus insist on the protection of private property in a traditional society dominated by communal rights, should they be called 'traitors' or are they contributing to the well-being of their community? No matter how we judge the choice of young members of traditional communities, we always use moral categories that say more about our own socioeconomic and cultural background than about the values and preferences of the traditional culture concerned.

2.5 Concluding Remarks Regarding the Theoretical Perspective

The 'polluter pays' principle is meant to guide the design of PES schemes which are aimed at internalizing the negative externalities of economic activities—but only in theory. In practice, PES tends to rely more on a 'beneficiary pays' or sometimes even a 'pay the polluter' principle. The inconsistency has a lot to do with PES being intended to solve too many problems with the same tool, but is also related to its theoretical baseline assumptions that are derived from neoclassical welfare economics. In this chapter we reviewed the development of PES in theory and practice and observed a trend towards hybridized approaches that are increasingly ignoring the narrow definition of PES. Instead more emphasis is given to the role of innovation and entrepreneurship in PES, designed to enable PES providers not just to passively benefit from a payment for an environmental service but also to enable them to make use of new knowledge and invest in the development of new sustainable solutions that can then be easily adopted by others thanks to the social network created through PES schemes.

The problem with welfare economics is that it is comparative-static in nature and therefore unable to recognize the welfare effects created by entrepreneurship and innovation. The economic model derived from it (representing an abstract world of two goods with different production costs and prices) is based on the implicit assumption that all goods and services that could possibly exist do already exist. The welfare generated through the introduction of new goods and services is thus largely ignored. This view stands in strong contrast to the reality of the global knowledge economy where new knowledge, technologies, products and services are constantly being introduced, resulting in new global markets with increasing returns (Warsh 2006; Romer 1994).

The increasing global accessibility of these goods has led to fast catch-up growth rates in the developing world thanks to the rapid adoption of new rules that have helped facilitate technology transfer, business development and the creation of new markets tailored to local demand. As the MEA report (MEA 2005) highlighted, this trend towards rapid economic change represents a huge challenge for the protection of environmental services, if governments remain passive or adopt policies that are ill-designed to cope with the challenges. The trend could, however, also represent a great opportunity to improve global ecosystem management, if governments assume leadership by creating an enabling environment that provides incentives to stakeholders in research institutions, the private sector and civil society to work together on new ways to make existing PES more effective and financially sustainable. At the same time, PES projects themselves could become a vehicle of change if they were able to create new forms of social capital on the ground, which go beyond like-minded groups and stakeholders, and if public sector regulation provides additional incentives to join PES schemes. Governments in emerging economies are successfully experimenting with such new systems of governance as has been illustrated by means of various concrete cases. This stands in strong contrast to the social planning approach that still prevails in the practice of PES in many affluent western countries, where farmers tend to be regarded as custodians of the environment and landscape beauty rather than people involved in business activities. Since farmers tend to be remunerated for playing this role, they are hardly likely to challenge this view, which prevails mainly among the non-farming majority of the population. However, even policy makers in affluent countries have started to realize that it is the acquisition of non-rival goods (knowledge/rules in the form of instructions, protocols, formulas, and designs that allow for the creation of new products and services through the process of innovation) rather than trade in existing rival goods (commodities) that is key to economic empowerment in rural areas. They have also recognized that such a transformation cannot be achieved by the private sector alone but must be facilitated by responsive institutions in government and higher education institutions and involve civil society (Ciolos 2011). All these insights are gradually being recognized in the PES literature as this chapter has illustrated.

By hybridizing and complementing classical PES schemes with new instruments, new collaborative approaches and appropriate technologies, public and private PES facilitators, especially in developing countries in need of sustainable change, aim at converting largely externally funded PES projects into markets for environmental goods that generate positive externalities for the community and the environment while also creating opportunities for local business that responds to local needs. These moves towards more financially sustainable and innovative hybrid PES schemes that take entrepreneurship and innovation into account are partially reflected in the PES trend towards landscape approaches and territorial development strategies in developed countries—and they have already become common in many developing countries that are forced to explore new ways to address the trade-offs between economic development and environmental sustainability.

Chapter 3
The Practical Perspective of Environmental Services Management: Field Studies on Innovation and the Remuneration of Positive Externalities in Agriculture in Kenya

Abstract The chapter illustrates the experience with different types of PES projects in Kenya. It starts with the Kenya Agricultural Carbon Project (KACP), a PES project designed to make use of voluntary carbon market to compensate farmers for carbon sequestration. Subsequently, the experiences with small-scale and large-scale watershed PES projects are discussed. Finally, the potential of entrepreneurship and innovation to create markets for environmental goods in Kenya is explored by means of selected case studies (Maassai-owned slaughterhouse, microinsurance for farmers and flower exports). All the case studies illustrate that the presence of the private sector is crucial in making use of PES as a vehicle to create a market for environmental goods.

3.1 Introduction

Kenya is widely considered to be a pioneer of innovative ways of remunerating farmers and pastoralists for the positive externalities they produce through the adoption of sustainable practices in crop and livestock production. In particular in the area of sustainable watershed management (which is also linked to carbon sequestration and biodiversity conservation), the government of Kenya has actively encouraged the development of PES schemes through the enactment of the Water Act 2002. This Act provided a framework for managing water resources and the provision of water services in Kenya. In addition, there are numerous public–private partnerships in different regions of Kenya that are committed to the improvement of ecosystem services by experimenting with different institutional designs that remunerate farmers for the positive externalities they generate for society and the natural environment. Kenya also has outstanding national research institutes, such as the Kenya Agricultural Research Institute (KARI), NGOs (e.g. Green Belt Movement), environmental consultancies (e.g. WAJIBU), and many universities that are engaged in applied research activities that support the various public–private partnerships in improving sustainable agriculture and rural

P. Aerni, *The Sustainable Provision of Environmental Services*, CSR, Sustainability, Ethics & Governance, DOI 10.1007/978-3-319-19345-8_3

livelihoods. Moreover, many of the leading international organizations working on the improvement of practices and policies related to environmental services in agriculture and livestock management are based in Kenya. They are also actively involved in the ongoing PES-related activities in Kenya.[1] Finally, Kenya is also well known for its creative use of user-friendly new technologies and the successful application of innovation in efforts to reduce poverty, empower people and protect the environment. The experimental approach pursued in Kenya to cope with the great national socioeconomic and environmental challenges of the twenty-first century is well in line with the main objective of the Kenya Vision 2030 (launched on 10 June 2008 by President Mwai Kibaki) to help transform Kenya into a 'middle-income country providing a high quality life to all its citizens by the year 2030'.

For all these reasons we chose to take a closer look at the PES-related activities in the country and at the role of innovation in the creation of positive externalities in agriculture. The in-depth case studies described in this chapter are based on field visits and numerous interviews conducted with experts in Kenya between August 27 and September 7, 2012.

The chapter starts with the assessment of the Kenya Agricultural Carbon Project (KACP) located in Kisumu in Western Kenya. KACP is a PES project that aims to take advantage of the emerging voluntary carbon market that compensates farmers for carbon sequestration provided that they are able to meet the necessary criteria. The discussion then turns to the challenges and opportunities related to the creation of small-scale and large-scale watershed PES projects, taking the cases of Lake Naivasha (small-scale) and Upper Tana (large scale). Finally, the role of innovation in the improvement of rural livelihood and environmental services will be discussed by means of two examples: the Kiserian slaughterhouse business and the Kilimo Salama plus project that sells microinsurance to small-scale farmers. Innovation also plays a major role in efforts to improve the sustainability of the export-oriented, private-sector driven aspect of Kenyan agriculture. In this context, we show how the concern for the public image and the requirement for compliance with strict private standards in the flower export business have substantially contributed to the sustainable innovation in the private sector. All the case studies illustrate that the presence of the private sector is crucial in making use of PES as a vehicle to create a market for environmental goods.

[1] These organizations include for example the International Livestock Research Institute (ILRI), International Centre of Insect Physiology and Ecology (ICIPE), World Agroforestry Centre (ICRAF), and the United Nations Environment Programme (UNEP).

3.2 Kenya Agricultural Carbon Project

Carbon Sequestration

Kenya Agricultural Carbon Project (KACP)

Project Milestones: *initiation: 2007, start of project implementation 2009, first payment due in 2012*

Environmental Service: *Climate Change Mitigation through the adoption of Sustainable Agricultural Land Management Technologies (SALM) in Kitale and Kisumu (60'000 farmers on 45'000 hectares)*
> SALM methodology was granted approval by the Verified Carbon Standard (VCS) in 2011.

Main Stakeholders: *World Bank (main buyer), Vi-Agroforestry (PES facilitator), ~20'000 small-scale farmers in Western Kenya in 2012 (main sellers)*

Remuneration of Positive Externalities: *Projected Carbon Payment: ~US$30/year/hectare + additional revenues from productivity enhancing practices and technologies and the sale of tree seedlings*

Source of Funding: *- Payment from Biocarbon Fund (World Bank): ~ US$ 2.1 billion in total*
- Project Costs since 2009 (Swedish International Development Agency (SIDA), CBOs) : ~US$ 500'000

Contact: *Bo Lager (bo.lager@viafp.org), Wilson Odongo (wilson.nyariwo@viafp.org)*

3.2.1 *Introduction*

The KACP was initiated in 2007 with the purpose of allowing small-scale farmers in the Nyanza region of Western Kenya to participate in the global carbon market through the promotion of sustainable agricultural practices that also lead to quantifiable amounts of carbon sequestration. The watersheds in the Nyanza region are characterized by unsustainable small-scale subsistence farming that has contributed to significant soil erosion and to pollution of the water of Lake Victoria. Low yields and market integration in agriculture also led to a high dependence of the region on food imports.

3.2.2 *Benefiting from Carbon Sequestration*

The method for carbon sequestration in soil developed by Vi-Agroforestry, a programme of the Swedish Cooperative Centre, together with the World Bank gained approval in 2011. KACP is the first agricultural project in Africa to calculate carbon sequestration in the ground using climate-adapted agricultural methods (SALM Methodology). Around 20,000 small-scale farmers in Western Kenya (average farm size <1 ha) agreed to become part of the carbon sequestration project in 2012. Once they have adopted SALM technology and are in compliance with the verified carbon standard, they will be eligible for payments that remunerate them for their contribution to climate change mitigation. Apart from agroforestry, the project promotes soil fertility and productivity-enhancing farm practices that also make producers less vulnerable to biotic and abiotic stress factors. Thus the introduction of SALM technology helps farmers to adapt better to climate change, to decrease water pollution, to prevent soil degradation in the region (see Box 1) and to improve on-farm food security.

Many farmers decided to join the initiative not only because of the advantages of sustainable land use management, but also because it gave them the opportunity to run a tree nursery business with the support of Vi-Agroforestry, selling tree seedlings to various local organizations, the Green Belt Movement and other farmers. The market for tree seedlings, however, is mainly driven by poverty reduction and environmental initiatives, whereas private sector investment has so far lagged behind.

A further incentive to become part of KACP is that local farmers benefit from the farm enterprise approach of Vi-Agroforestry promoting market linkages and access to small loans for small enterprise groups of producers. Enterprise activities also include biogas, solar power and biodiesel production, which may also reduce wood consumption and thus deforestation. The material value of these additional farm benefits, which are unrelated to the payment for carbon sequestration, may be the true contribution to dealing with the challenging socioeconomic conditions in the region (see Box 2).

Even though the introduction of these new goods and services by Vi-Agroforestry is generating positive externalities for society and the environment in the region, it is unclear whether they can be sustained without the generous funding from Sweden.

BOX 2

Agro-ecological conditions and farm activities

The agro-ecological conditions in the region of Kisumu are characterized by low soil organic matter resulting from continuous cultivation with negative nutrient fluxes. This form of land degradation is a result of the harmful situation of population growth without structural change. The smaller the farms become, the more intensively the small plots are cultivated, and the poorer the farmers are the less capacity they have to replenish the soil. There is a high rate of pest infestation. Serious pests include leaf beetles, aphids, thrips, white flies, mealy bugs) and plant diseases, such as the cassava brown streak and mosaic viruses have contributed to the harvest losses in the whole of the Lake Victoria region. There is also a high incidence of abiotic stress factors such as drought and flood (also related to lack of drainage systems). The floods, in particular, are related to the high degree of deforestation, encroachment of wetlands by agriculture and industry, and the expansion of small-scale farming over the past 30 years (using wood as the main source of energy and clearing forest to expose more fertile soil). The number of small-scale farms in the Nzoia Watershed near Kisumu increased from 36,292 in 1986 to 263,385 in 1995 and remains high. The intensive use of poorly drained and acid soils with high clay content leads to a high level of rainwater surface runoff and the siltation of Lake Victoria which is being ever more contaminated: the persistence of the water hyacinth pest in the lake is the most visible indicator. In addition to the problem of lack of facilities for treatment of waste water from urban settlements and industry, the physical and chemical analysis of water quality reveals high levels of phosphates and nitrates along the agricultural zones of the River Nzoia Basin. The in-situ measurements of the physiochemical parameters of water indicate that values are generally higher than the United States Public Health Standards. The higher values of electrical conductivity from in-situ tests are an indicator of severe soil erosion resulting from land degradation. Soils from some selected fields in the Nzoia River basin showed high levels of illegal compounds such as aldrin, dieldrin, endosulfan, DDT, and endrin, which are referred to as persistent organic pollutants (POPs). Despite the detection of illegal crop protection compounds, farming in the Kisumu region is generally characterized as low-input/low-yield agriculture that mainly serves subsistence purposes. Average corn yield per hectare in this region is 1 ton compared to 4–5 tons in neighbouring regions. The lack of market infrastructure also helps explain the high proportion of postharvest losses (lack of: storage facilities, dependable road network, integration into the formal food chain, and commercially-oriented cooperatives). Numerous organizations are taking action in the region to amend the situation through more sustainable agronomic practices; improved management of nutrients, water, tillage and residues; agroforestry; rehabilitation of degraded land; and integrated livestock management. Yet, the situation has not changed much over the past decade if the water hyacinth is taken as an indicator. People say the lake is dying despite millions of US dollars that are invested every year by national and multilateral donors to save Lake Victoria.

With regard to the benefits of participating in the global carbon market, the project also faces serious challenges that cannot be addressed by the project managers themselves. The costs of promoting and maintaining climate-smart agricultural practices, and the huge transaction costs involved to complying with the

Verified Carbon Standard (VCR) mean that it is not a very attractive investment opportunity for the private sector (including farmers).

Even though the aggregate level of sequestration that could be achieved at relatively low costs by increasing carbon in soils is considerable, there is still relatively little field experience with crediting mitigation from soil carbon sequestration. So far only costly project-based approaches (such as KACP) to generate offsets for the voluntary market have been created.

BOX 3

Socioeconomic conditions

The region in Kenya where KACP is being implemented is characterized by a high incidence of food poverty (53.4% compared to a national average of 29%), above average birth rates (an average of 5.6 children per woman) and small average farm sizes (0.8 hectares). Even though the level of education is high, there is a serious problem of youth unemployment (general unemployment rate is above 30%). Cities are usually the engines of off-farm employment creation. They indirectly enable structural change in the countryside by bringing knowledge and technology to rural producers and offering markets and employment for migrants. However, the city of Kisumu offers few off-farm employment opportunities in the formal sector outside government and civil society institutions (uncompetitive sugar and cotton industries, a few tourism activities). The small average farm size, low degree of farmer organization and poor infrastructure in the countryside explain why most food consumed in Kisumu is actually imported from other districts rather than being bought from the surrounding subsistence farmers. Unlike in typical watersheds, upstream dwellers seem to be better off than downstream inhabitants in the watersheds of Western Kenya. Owing to the differences between the downstream tribe (Luo) and the upstream tribes (mostly Kalenjin and Kipsigis), there may also be a lack of the social bonds and trust that would otherwise easily facilitate voluntary agreements. Overall, these factors make a classical PES scheme with downstream buyers and upstream sellers of environmental services difficult.

The voluntary markets were mainly created in anticipation of their eventual acceptance into compliance markets. But what if such compliance markets remain elusive? The current low prices for soil carbon offsets in agriculture, forestry and other land use (AFOLU) reflect the low expectations regarding a functioning global carbon market and they make it difficult to capture agricultural mitigation benefits in developing countries. The voluntary carbon market in which KACP participates mainly trades in verified emissions reductions (VER). These, however, cannot be used for regulatory compliance for emissions trading under the Kyoto Protocol. Eligibility of agricultural land use offsets is extremely limited in the major compliance markets. For example, the CDM only recognizes a limited range of smallholder and agriculture-relevant activities, and agricultural soil carbon is not one of them. The EU-ETS does not allow trade in land-use offsets, even if they relate to afforestation and reforestation activities that are eligible for the CDM.

At the moment, it is only the accord on Reducing Emissions from Deforestation and Forest Degradation (REDD+) that offers hope of becoming a source of carbon income for farmers who participate in the voluntary carbon market (see Sect. 1.3.1).

The uncertainty of the global carbon market (especially within the context of soil carbon sequestration in agriculture) is making it hard to predict to what extent farmers will financially benefit from soil carbon sequestration.

3.2.3 Lessons Learned

The KACP shows that the current potential for farmers to benefit from payments for the delivery of carbon sequestration services is very limited. Payments for this service are likely to remain very low in comparison to the overall costs of complying with applied voluntary carbon standards. Moreover, farmers, as the sellers of the environmental service, are unlikely to benefit in any other way from the non-resident single buyer, the World Bank's Biocarbon Fund. The fund has no roots in the region as an economic actor that could reward farmers who embrace CSA through other forms of in-kind payments or access to appropriate technology or new markets. Instead it is Vi-Agroforestry, the facilitator of the PES scheme, that has stepped in to provide sufficient additional incentives for farmers to make it attractive to join the initiative. The additional benefits comprise productivity increases through improved agricultural practices, cash revenues through the sale of tree seedlings and improved access to sustainable input (for example, alternative energy technology, loans, conservation plant material) and output markets for organized producer groups.

The main benefit of the KACP project is therefore that farmers are given opportunities to improve their livelihoods in a sustainable way. The main risk is that these livelihood improvements are not really contributing to meeting the need for structural change in local agriculture. The trends towards shrinking farm sizes, intensive farming without sufficient soil replenishment and low capacities to produce food surpluses and off-farm employment remain long-term risks to local food security and local environmental sustainability (see Box 1), especially in view of the potential reversibility of the positive externalities achieved with the support of Vi-Agroforestry. Vi-Agroforestry is aware of this challenge and invests a significant amount of its budget in the market integration of farmer groups.

3.3 Small-Scale Watershed Project: The Case of Lake Naivasha

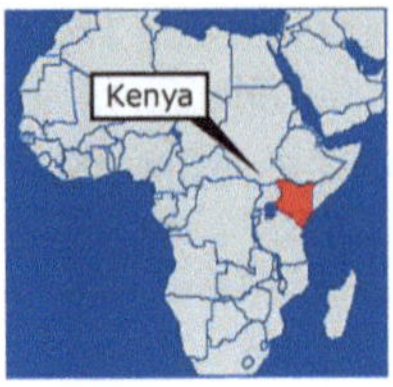

Water Quality Service
-Small-Scale -

Naivasha Malewa
Sub-basin Project

Project Milestones: *2006-07 (feasibility studies), 2008-11 (piloting), 2011-(upscale)*

Environmental Service*: The sellers adopt land use changes that results in provisioning of good quality/clean water (free from sediments and other fertilizers) to the Lake,*

Main Stakeholders: *Private Sector Naivasha(main buyer), WWF/CARE (PES facilitators), 785 small-scale farmers in the NaivashaMalewaWatershed in 2012 -represented by their respective WRUAs (Water Resource Use Association) as the seller vehicle.*

Remuneration of Positive Externalities: *Payment for Watershed Services: ~US$70 per farm household paid out in vouchers. Additional revenues from productivity enhancing practices*

Source of Funding: *- WWF/CARE costs of setting up and maintaining the PES scheme (~US$ 120'000)*
- Annual Payments from the local private sector (~US$ 10'000)

Contact: *Peter Muigai (mpmuigai@gmail.com), Josephat Nyongesa [nyongesajm@yahoo.com)*

3.3.1 Introduction

Payments for watershed functions, such as water quality and quantity, seek to link upstream land use and management with downstream water use to realize social and environmental benefits for upstream and downstream participants. They take the form of a voluntary agreement between at least one buyer and one seller of ecosystem services (or land-use changes presumed to provide an ecosystem service). The Malewa-Naivasha Sub-basin Watershed Project can be considered a model case of a small-scale PES project in watershed management where payments have served so far as a vehicle for change in the region, offering small-scale farmers new economic opportunities, provided that they embrace sustainable agricultural practices and thus contribute to the reduction of river siltation and to the improvement of the water quality in the region. Their activities benefit not only the water users downstream around Lake Naivasha, thanks to better water quality, but also society, the economy and the environment in the region as a whole.

Even though the PES project is still far from achieving its full potential (roughly a quarter of the upstream farmers have so far enrolled in the scheme), there are signs that the project initiators, WWF and the Cooperative for Assistance and Relief Everywhere (CARE), may eventually be able to withdraw from the project without running the risk that the PES project will collapse owing to its dependence on outside support. CARE withdrew in 2011. Such an exit strategy is viable because PES has been used to bring local actors with different interests together and make them part of a voluntary agreement that seems to be beneficial to all the parties involved. In this context, the downstream private sector, and in particular the flower export business around Lake Naivasha, have not just become the main buyers of the services but they are likely to help convert PES into a local market for environmental goods that ensures financial sustainability by making the adoption of sustainable land management practices a condition for access to all kinds of additional benefits.

3.3.2 Project Background

Lake Naivasha is the only inland freshwater lake of economic importance in Kenya. The region is part of the rift valley, with the well-known natural beauty and wildlife that characterizes many of the semi-arid regions of Kenya (see also Box 3). It has been recognized as a 'wetland of international importance' under the Ramsar Convention on Wetlands. At the same time Naivasha supports flourishing businesses in tourism, horticulture and floriculture. The large flower export industry is a particular subject of controversy. It generates much employment but has also been criticized in the media for causing social and environmental problems in the region. There is no doubt that many of the flower producing companies did not initially care much about worker welfare or about the discharge of polluting effluents from the greenhouses. However, negative media coverage and the need for global retailers of flowers to comply with global private standards[2] has led to significant positive changes in recent years. Numerous chemicals that were deemed to be harmful to the environment, independent of the actual quantity in treatments, have been phased out and every flower company with an export licence must have its own wetlands to purify its effluents; ensure workers' health and environmental safety standards on site; and contribute to social welfare in one form or another. Even though compliance with these standards (enforced by the Kenya Flower Council) is very capital-intensive and thus makes it impossible to operate an outgrower scheme in which small-scale farmers could also participate, it has helped to improve the social and environmental record of the flower business. Even WWF admits that the main source of pollution of Lake Naivasha is not the flower industry, but the numerous small-scale farmers upstream that are unable to comply with capital-intensive private standards; tend to apply means of plant protection in an inappropriate way; and use pesticides that have been officially banned. Another problem is that they tend to cultivate slopes while taking few soil conservation measures, causing soil erosion, river siltation and

[2] See http://www.fairflowersfairplants.com/home-en/ffp-standard.aspx

ultimately water eutrophication in Lake Naivasha where the emergence of the water hyacinth is one of the main indicators (see Box 3).

3.3.3 History of the PES Project

WWF and CARE recognized in 2006 that the River Malewa Basin was threatened by the encroachment of farming activities into riparian areas and by unsustainable small-scale farming practices resulting in deforestation, siltation and water pollution (see Box 3). As a result, the water output of the river decreased and water quality was compromised. This situation constituted a threat to the flourishing economic activities along the shores of the lake as well as to the environmental health of the river basin and of Lake Naivasha. Improved watershed management had been identified as crucial for improving water quality in the catchment, but there were no incentives for upstream farmers to contribute to it through the adoption of sustainable agricultural practices.

BOX 4

Agro-ecological conditions

Lake Naivasha is supplied by three rivers and has no surface outlets other than underground outflows. River Malewa, the main river, rises in the western slopes of the Aberdare ranges in central Kenya and flows south through numerous farming areas before entering the lake on its northern shores.

The area around the lake is semi-arid and encompasses the PES pilot sites of the Wanjohi area and Upper Turasha-Kinja. The terrain varies from steep and moderately steep, to undulating downstream from the upper parts. The land use includes mixed farming dominated by potatoes, vegetables, flowers and dairy cattle. The area receives high rainfall, between 1,000 and 1,300 mm per annum, has fertile soils and is predominantly home to the Kikuyu community who settled there in the postcolonial period on parts of the former large-scale White Highlands farmlands. In recent decades, the large-scale wheat and dairy farms have been subdivided into small land parcels for subsistence farming. The average size of a farm today is 0.6 hectares. Owing to the overuse of natural resources and unsustainable farming practices in the upstream area, Lake Naivasha and the Malewa catchment are threatened by environmental degradation and loss of biodiversity.

At the same time, the private sector downstream was largely focused on compliance with its own social and environmental standards and was under too much short-term commercial pressure to also be able to take into account the long-term regional economic impact of the negative externalities resulting from water siltation and decreasing water quality. Yet, they were aware of the problem and willing to tackle it. These were the ideal conditions to make a PES scheme work since there were sufficient incentives to participate on the side of the downstream industry to act as main buyers and on the side of upstream farmers to become sellers of environmental services. The joint WWF-CARE project consisted of three phases. It started in 2006 with scoping and feasibility studies including a hydrological survey, a cost–benefit analysis, a livelihood analysis, a business case analysis and a legal policy framework analysis. The pilot phase started in 2008 and ended in 2011. The first payments were made in 2010 to 470 farmers in the form of vouchers. In this context, the local branches of the Water Resource Users Associations (WRUAs) provides Certification through on farm verification and hydrology monitoring.[3]

In 2011, the project entered its third and final phase—the scaling up phase. One of the main purposes of this phase was to enhance the participation rate of upstream farmers (784 in 2012) so that the improvements in the quality of the river water would become measurable.

The WWF-CARE project represents a classic PES scheme in sustainable watershed management. The sellers of the service 'water quality' are upstream farmers who adopt land use changes that result in the provision of good quality/clean water (free from sediments and fertilizers) to the lake area. The main buyers are represented by the private sector downstream. As important water users, the private sector companies are also considered to be the main beneficiaries of the service. The private sector consists not only of the flower business but also includes water companies, horticultural growers, ranchers and the hotel and tourism industry. WWF does not act directly as the intermediary between downstream buyers and upstream sellers but instead helps the local WRUAs, representing both parties, to reach voluntary agreements (see Box 4).

[3] For more details, see http://www.fao.org/fileadmin/user_upload/pes-project/docs/FAO_RPE-PES_WWF-Kenya.pdf

BOX 5

Water resource users associations (WRUAs)

The WRUAs are managed by the local Water Resource Management Authority (WRMA). WRMAs were established all over the country with the creation of the Kenyan Water Act in 2002. The main purpose of the Water Act is to provide for the management, conservation, use and control of water resources and for the acquisition and regulation of rights to use water. It also requires the WRMAs to formulate a catchment management strategy for the management, use, development, conservation, protection and control of water resources within each catchment area. In this context, the WRMA must count on the support of WRUAs to help them measure the local water use and consumption, disseminate information on water conservation techniques, create awareness of the importance of sustainable water use and serve as a bridge between the WRMA and existing water projects. However, WRUAs have a funding problem: the WRMA may support them by reserving around 25% of the revenues they generate from selling water permits to water users downstream for water conservation projects upstream. But generally WRUAs are expected to fund themselves through joining and annual membership fees, which vary according to membership category. Such fees, however, are tiny or non-existent in poor areas that are largely dominated by small-scale subsistence farmers. Even though WRUAs are invited to apply for finance for particular projects with The Water Services Trust Fund that was also set up with the 2002 Water Act, most WRUAs have difficulty in gaining access to this fund. In response to the scarcity of funding, many WRUAs have also started to generate revenues by setting up businesses that help farmers to comply with sustainable practices, such as starting a tree nursery business.

3.3.4 The PES Scheme of Lake Naivasha

In the case of Lake Naivasha, it is the WRUAs that are in charge of the PES scheme. The seller WRUAs upstream engage in a contractual agreement with the LANAWRUA (Lake Naivasha WRUA), the buyer WRUA downstream, on an annual basis. The WRUAs upstream are in charge of on-farm verification. Only farmers who are found to have implemented the soil conservation measures qualify for the incentives. On-farm verification involves three types of monitoring:

(i) On-farm monitoring: the verification of the implementation of conservation measures by the seller WRUA is done in collaboration with the Ministry of

Agriculture (MoA). It ensures that the conservation measures are in line with the recommendations of the MoA. This on-farm monitoring also includes the measurement of accumulation of organic soil in the soil conservation sites of the field. The buyers also undertake the monitoring to check whether the farmers are actually implementing the measures before they release the incentives.

(ii) Socioeconomic monitoring: the project is also expected to improve the livelihoods of the participating communities. Monitoring is therefore done to evaluate the extent to which the project has improved livelihoods and this is mainly carried out by experts assisted by the WRUAs.

(iii) The impact of the project on water quality: water quality is the key environmental service to be delivered. The link between sustainable farming practices and improved water quality is examined through the hydrological monitoring (see Box 5). In this context, the project has facilitated the installation of four river gauging stations and sampling sites. WRUA members have been trained to collect data on river flows and turbidity, which they do daily. The data collected is then submitted to the Water Resource Management Authority (WRMA) for analysis. This process has been running for the past 2 years and so far the data available is not sufficient to enable the experts to draw firm conclusions concerning the water quality. Non-point-source sedimentation from degraded public land may largely account for this inconclusive result since such sedimentation may obscure the hydrological benefits arising from land-management improvements on the targeted hotspot farms. But it is also related to the fact that only about a quarter of all the farmers upstream are currently participating in the PES scheme.

BOX 6

Hydrological monitoring

The catchment of Lake Naivasha covers an area of approximately 3,400 km^2 and ranges in altitude from approximately 1,900 m to about 3,900 m above sea level. The project area covers the Turasha-Kinja and Wanjohi rivers. Both are major tributaries of the River Malewa, which contributes 80% of the water that flows into Lake Naivasha. To measure the impact of the project on water quality – the key environmental service, the project has facilitated the installation of four river gauging stations and sampling sites. WRUA members have been trained to collect data on river flows and turbidity, which they do daily. The data collected is then submitted to WRMA for analysis. This process has been running for the past two years and so far the data available is insufficient to enable the experts to draw firm conclusions concerning the water quality: thus it is expected to run for several more years. The report generated from this data will be shared among different stakeholders in the Basin and, most importantly, communicated to the buyers to prove the provisioning of the environmental services and encourage them to invest more in the scheme.

The goal is therefore to reach many more of the farmers and then run the hydrological monitoring over several years. The report generated from these data will be shared with the buyers to prove the provisioning of the environmental services and thus encourage them to invest more in the scheme. But this would require more resources for the cash-strapped upstream WRUAs.

The private sector downstream, as the main buyer of PES, is aware that environmental benefits obtained through the adoption of sustainable agricultural practices are of a long-term nature and that the improvements to rural livelihoods in the upstream region are also a positive externality for the entire region. Moreover, WWF provides an additional incentive to the private sector to act as buyers by offering them WWF certificates that confirm their participation in the PES scheme. Such certificates help companies to enhance their reputation as corporate citizens and they help WWF to ensure the financial sustainability of its PES project.

3.3.5 WWF as a Potential Facilitator of Local Markets for Environmental Goods

WWF spends a considerable amount of money (via WRUAs) to support upstream farmers in their efforts to embrace sustainable agricultural management practices.

Upstream WRUAs are paid to act as local intermediaries, community organizers and providers of environmental services to farmers. The products and services they offer to farmers comprise:

- grass strips to check soil erosion and act as filter materials. Napier grass (*Pennisetum purpureum*), cocksfoot (*Dactylis glomerata*) and Rhodes grass (*Chloris gayana*) used for conservation, have increased fodder supply resulting in higher milk production and reduced pressure on forests from grazing.
- reduced use of fertilizers and pesticides, e.g. through integrated crop and pest management, and the use of new or improved crop varieties.
- agro forestry along the grass strips—this is to reinforce the grass strips and improve their effectiveness in erosion control. Planting fruit trees and the use of higher quality material for potato planting bring in additional income and provide more nutritious food.
- support for river bank/riparian rehabilitation by planting grass and water-friendly trees along the riparian areas to act as a buffer. Such support is essential for farmers that have settled in riparian areas and were thus in conflict with the Water Act that generally requires riparian areas to be free of agricultural activities.
- good land management practices, such as cultivating along the contours as opposed to across the contours, crop rotation and other agro-ecological land management techniques to maintain soil fertility and ensure soil conservation.

WWF buys the conservation plant material (tree or plant seedlings) from the Kenya Agricultural Research Institute (KARI) and funds its transport to the WRUAs upstream where it is subsequently given to farmers for free. Considering that the transportation costs are almost as a high as the price of the plant material, it would probably make more sense to encourage the WRUAs to develop their own nursery business. This would allow them to generate additional revenues and save WWF the costs of funding the transport.

Interviews with the manager of a dairy cooperative in the upstream region revealed that farmer cooperatives might be very interested in promoting sustainable agricultural practices among their members by offering them special discounts on inputs or lower interest rates for loans in return for embracing sustainable agricultural practices. These cooperatives are operating in a competitive environment and are well integrated into the national dairy value-chain. The more members they have, the more likely they are to be in a position to fix a decent price for their milk in the negotiations with the large dairy companies that operate in Kenya. To make the cooperative attractive for farmers, the management has developed all kind of services for their members. For example, the cooperative picks up the milk at the farm gate and ensures sustainable livestock management and quality control through numerous extension workers. The manager affirmed that it would be feasible for such extension workers to also monitor the introduction and maintenance of sustainable agricultural practices in return for input discounts.

3.3.6 The Need for Structural Change to Make Farming Sustainable in the Long Run

Interviews with managers of the flower business in the Lake Naivasha region revealed a willingness to go beyond PES alone by offering to source food products for their canteens directly from upstream farms that have been certified as sustainable agricultural producers. In addition to paying farmers a small price premium the flower businesses could offer them a stable local market for their food products. This market is considerable because the Naivasha flower business employs more than 6000 people. A similar market could be developed in the Naivasha tourism industry. Hotels could source the food from selected upstream farms that serve as models for sustainable agriculture and offer guests the chance to participate in agro-tourism activities by arranging excursions to the farms that produce the food for the hotel. It is therefore the presence of a strong private sector that could help to convert PES into a market for environmental goods. The private sector also plays a crucial role as an engine for the generation of off-farm employment, which is crucial not only for the economic but also for the environmental sustainability of the region. Considering that the average farm size in the upstream regions is less than 0.6 ha, there is an urgent need for sustainable structural change (see Box 6). Farm households that are able to pay the school fees for their children may hope that one day these children will find work outside the farm, because otherwise the farm would have to be subdivided into even smaller plots. Smaller farm sizes, however, may no longer be sustainable, either from an economic or from an environmental point of view. The pressure to squeeze out sufficient food from the remaining small plot with ever less means to replenish the soil will lead to greater food insecurity and environmental degradation.

If the pull-factor of the private sector expansion is sufficiently strong, however, a higher proportion of the educated offspring of farmers will find work outside agriculture, which in turn will help to improve the livelihood of the parents back on the farm through remittances. At the same time, this would allow the remaining farms to grow in size. At a later stage, the move to urban settlements will lead to an increase in forest area as has been observed in many developed countries, and recently also in emerging countries, that had passed through the first stages of structural change in agriculture. Ultimately, environmental protection will then be largely focused on ecological restoration and conservation activities outside agriculture. In turn, farmers themselves will have a genuine interest in complying with the standards of sustainable intensification, not just because of the payments they would receive for their environmental services but also in view of the increased long-term benefits from farming itself.

Looking at PES from this long-term perspective would allow WWF to exit the PES project in Naivasha without endangering its financial sustainability. It could then focus instead on other activities in the area that improve biodiversity, wildlife and the aquatic environment.

BOX 7

Socioeconomic conditions

The land in the upper catchment of the Malewa River was forest before the colonial era. The colonialists who settled in the region started dairy and wheat farming. After independence, the land was allocated to the Kenyans who started clearing the vegetation in order to create land for agriculture and livestock rearing. The area is adjacent to a gazetted forest land which also hosts the Abardare National park – a major tourist attraction in Kenya. Kenyans who settled in the region after independence acquired the title deeds, and land subdivision began. The population has been increasing steadily, creating a high demand for agricultural land and leading to massive deforestation and degradation of the natural resources. The average farm size in the upstream areas is around 0.6 hectares. The land ownership is private, with the majority of the landowners having title deeds. About 5% of community is engaged in small-scale businesses, with artisanal occupations and transport being the main ones. The largest shares of off-farm employment, however, are generated by the flower export business and tourism along the shore of Lake Naivasha. The flower business alone employs more than 6000 people.

3.3.7 PES as a Vehicle to Create a Market for Environmental Goods

As with the KACP, it is therefore not the payment as such that is the real incentive for farmers to participate, but the additional benefits they gain from being part of the scheme. In the case of the KACP these comprise the farm revenues generated from the tree nursery business, access to loans and sustainable technologies (e.g. solar lamps, biogas)—in addition to the productivity increases obtained through the adoption of sustainable land management practices. However, the problem in the Kisumu region is that the buyer of the service of soil carbon sequestration (biocarbon fund of the World Bank) is not a local player that could help with generating markets for environmental goods. Instead it is Vi-Agroforestry, the initiator and intermediary of the project that serves as the main provider of loans, services and technologies and as the buyer of farmers' goods. Even though these activities help to improve the livelihoods of farmers, this situation makes it difficult for Vi-Agroforestry to exit by creating a market for environmental goods. Moreover, in view of the absence of a vibrant private sector and, consequently, the limited off-farm employment opportunities, structural change is unlikely to prevent farms from shrinking to a size that will be unsustainable. In this context, the small-

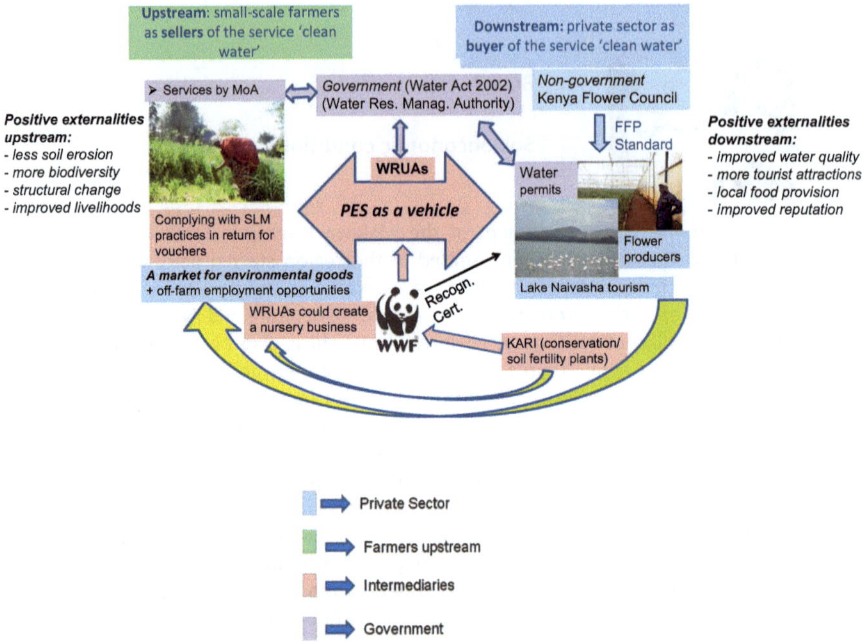

Fig. 3.1 The PES Scheme of Lake Navaisha

scale PES scheme in Naivasha may have greater potential to succeed in the long run, not because the payments for farmers are significantly higher but because of the existence of a private sector that offers all kinds of in-kind payments and market opportunities that could eventually help to establish a market for environmental goods that generates large positive externalities for the region as a whole.

Figure 3.1 shows how the initial idea of implementing a classic PES scheme in Lake Naivasha has turned into a hybridized and multilayered institutional arrangement that benefits buyers and sellers in multiple ways and is also supported by the country's existing environmental law, especially the Water Act of 2002. The figure illustrates the different interactions in this hybridized PES scheme highlighting the role of the various stakeholders (private sector, civil society, government, academia) in different colours.

3.4 Large-Scale Watershed Project: The Case of Upper Tana Basin

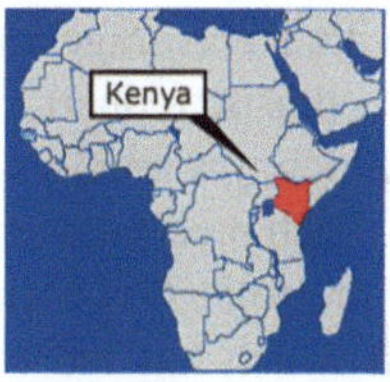

Water Quality Service

- Large-Scale -

Upper Tana Basin

Project Milestones: *Mount Kenya East Pilot Project (MKEPP) initiated in 2004, Upper Tana Catchment Natural Resources Management Project (UTaNRMP) approved in 2012 (upscaling) + Water Fund of The Nature Conservancy created in 2012*

Environmental Service: *Adoption of Soil and Water Conservation (SWC) -in return for livelihood support- in the Upper Tana Basin providing water users downstream with adequate water quality and quantity*

Main Stakeholders: *Private Sector Downstream (main buyer), The Nature Conservancy (TNC) (PES facilitator), 205'000 poor rural households - represented by 240 WRUAs (Water Resource Use Association) as the seller vehicle.*

Remuneration of Positive Externalities: *Creating a Water Fund/ Green Water Credits to reward those who implement sustainable farming practices and create benefits for environment and society*

Sources of Funding: *- Global Environmental Facility (GEF), International Fund for Agricultural Development (IFAD), US Agency for International Development (USAID), Spanish Trust Fund, Government of Kenya, The Nature Conservancy million*

-

Contact: *Fred Kihara (fkihara@TNC.ORG), Boniface Mwaniki [bm.mwaniki@yahoo.com)*

3.4.1 Introduction

The Upper Tana catchment covers an area of 17,420 km^2 and includes 24 river basins that drain into the 1000 km long Tana River. It includes the Mount Kenya and Aberdare national parks and surrounding forest reserves. The area covers six of Kenya's 47 counties, is home to 5.2 million people and provides water for about half the population. It is the primary water supply source for people living in the vibrant capital city of Nairobi. The basin encompasses diverse agro-ecological zones. Its agricultural potential and production therefore varies from one catchment to another. Coffee and tea are the main cash crops grown in the upper part of the basin. Horticulture, flower and maize production are present in the high potential areas of Upper Tana. Millet, cotton, tobacco and sorghum are more widespread in the low potential areas. There are also large numbers of livestock producers and a well-developed dairy business. Nearly 80 % of the basin is arid or semi-arid and around 50,000 ha are irrigated. Five major dams along the midsection of the basin provide water for irrigation and most of Kenya's power through hydroelectric generation. The lower parts of the Tana catchment are dominated by extensive rangelands. Important wildlife conservation sites with several game parks and tourist destinations are found in almost all sections of the Basin. The expansion of industrial and commercial activities and the changes in land use practices in the Basin have increased the demand for various natural resources. At the same time, the area is under heavy and growing population pressure, with an average of 300 inhabitants per km^2. The catchment has been experiencing a drastic reduction of surface water availability especially during the dry season, which is a manifestation of high runoff rates and decreasing groundwater recharge. High precipitation in the upper recharge area also causes floods (over 70 % of the total flow). In addition, the area has experienced increasing land degradation which has led to siltation and pollution of the rivers and dams. The daily sediment load of the Tana River varies from 2800 tonnes/day during the dry season to more than 24,000 tonnes/day during the rainy season.

In view of the economic and environmental importance of the Upper Tana Basin, several national and international initiatives have been set up to make water use more sustainable and equitable by supporting farm activities and off-farm activities that produce positive externalities for society and the environment in the region. Two international initiatives to improve ecosystem services and livelihoods in the region are being implemented: the Upper Tana Catchment Natural Resource Management Project (UTaNRMP), which is the upscaling phase of the earlier Mount Kenya East Pilot Project (MKEPP), and the Upper Tana-Nairobi Water Fund set up by The Nature Conservancy (TNC). Both projects include a PES/RPE component but they are also different from classic PES schemes. The UTaNRMP focuses on improvements to livelihood as a means to introduce more sustainable soil and water conservation (SWC). The TNC project aims instead at leveraging public and private sector investment through the creation of a water fund that pays for the protection of water quality and quantity. Both initiatives are considered to be supportive of the main objective of ‘Vision 2030’ established in 2008 by the Kenyan Government to bring about greater and more sustainable growth of the economy in a more equitable environment, accompanied by increased employment opportunities. Agriculture, livestock and fishing is one of six priority sectors expected to deliver 10 % annual growth.

The numerous PES-related initiatives of the Upper Tana Basin illustrate the coordination challenges related to the management of a large-scale water basin with multiple watersheds. In the case of Lake Naivasha, buyers, intermediaries and sellers are comparatively few and the region seems manageable in terms of socio-economic and agro-ecological conditions. By contrast, the diversity and interconnectedness of these elements in the Upper Tana Basin seem to require case-by-case approaches that are subsequently coordinated on a regional and ultimately on the national level, reflecting the importance of the basin for the provision of water and electricity to the capital, Nairobi. This also helps explain why PES/RPE projects have not reached the maturity of that of Lake Naivasha. The exploration of approaches with pro-poor rewards for environmental services and Green Water Credits (GWC) have not gone beyond an exploratory phase in during the MKEPP phase. Even though there were assurances that GWC will be part of a PES scheme in the UTaNRMP and TNC projects, there seems to be a reluctance to commit to any one particular approach to remunerating farmers for the positive externalities they might generate in the water basin by adopting sustainable agricultural practices. This is not have to be necessarily a disadvantage, since there is no consensus about the most appropriate RPE scheme or whether a comprehensive PES scheme would be viable in a water basin the size of Upper Tana. Experimentation with different approaches within different sub-catchments of the Basin might therefore be more useful and less risky than relying on one particular approach that would involve all types of watersheds regardless of their diverse interest groups and agro-ecological conditions.

3.4.2 Upper Tana Catchment Natural Resource Management Project

The MKEPP started in 2004 with the assistance of International Fund for Agricultural Development (IFAD). Its aim was to link sustainable use of natural resources, especially water and forests, with enhanced rural livelihoods. Associated with the MKEPP-IFAD loan were two IFAD grants with a link to PES and GWC. Pro-poor Rewards for Environmental Services in Africa (PRESA) has been exploring opportunities to develop systems of payments or rewards for provision of ecosystem services (RES) that not only address environmental challenges but can also improve the livelihood of farmers. GWC has focused instead on identifying SWC measures for smallholders in the Upper Tana catchment and the institutions required for the successful implementation of the scheme. GWC is therefore a special type of PES in the sense that it typically focuses on the upstream/downstream interaction and interdependency within a watershed (see Box 7). Neither approach, however, has so far gone beyond feasibility studies that provide proof-of-concept for the next funding phase. This next phase is the UTaNRMP. It aims to extend the reach of MKEPP from 5 to 19 river basins in the Upper Tana Basin and will run for 5 years (2012–2017) with the support of GEF, IFAD, USAID, the Spanish Water Fund and The Government of Kenya (IFAD 2012).

BOX 8

Green water credits

Run-off and high evaporation rates lead to the loss of green water, defined here as the water that is held in the soil. Soil and water conservation measures such as mulching and contour cropping reduce the amount of runoff and evaporation. As a result, the amount of water available for plants and deep percolation to the groundwater (blue water) increases.
Green water credits (GWC) provide a mechanism for transfer of cash or other benefits to rural people in return for water management activities that determine the supply of green and blue water at source. The online database of WOCAT (World Overview of Conservation Approaches and Technologies) illustrates that there are thousands of farmers already applying land and water conservation practices whose services are currently unrecognized. Feasibility studies to document the GWC proof-of-concept were supported by the International Fund for Agricultural Development and the Swiss Agency for Development and Cooperation in 2007.

The scaling up of MKEPP through UTaNRMP will be based largely on the aspects that worked well in MKEPP, with appropriate modifications based on lessons learned. In the upscaling phase, greater emphasis will be placed on reducing human–wildlife conflicts which jeopardize the conservation-livelihoods thrust of the project, as well as on increased involvement of private sector and research institutions in the development of public–private partnerships for more sustainable agriculture.

Management of water resources will be based on the river basin management concept that was piloted successfully under MKEPP, taking into consideration the decentralization of administration. Based on the experience of implementation of MKEPP and other relevant NRM projects, the basic approaches to the implementation of UTaNRMP will focus on a two-pronged strategy that addresses poverty and natural resource management simultaneously, on the basis that they are closely interlinked. This means that all natural resource users in the catchment area will benefit in some way. However, there will be special measures to ensure the inclusion of disadvantaged and vulnerable groups, without excluding the better-off households. The specific mechanisms to ensure that the poor benefit from the project interventions take into consideration the population, poverty levels, the ecosystem, gender and other social issues. Yet, there seems to be a reluctance to adopt a particular PES scheme to reward farmers for the positive externalities they generate for society and the environment.

3.4.2.1 Rewards for Environmental Services

Initially, UTaNRMP intended to set up a financial compensation scheme that would allow farmers to obtain loans under favourable conditions in return for adopting specified Soil and Water Conservation (SWC) practices. For that purpose, a USD 10 million credit guarantee fund was envisaged to be accessible by a commercial bank as an incentive to advance loans.[4] The water and electricity companies were also expected to contribute to the guarantee fund as a form of PES. This conditional approach to the provision of financial services was dropped out of concern that it would lead to the duplication of already existing activities in the region. For example, the IFAD/Alliance for a Green Revolution in Africa (AGRA) is supporting a Programme for Rural Outreach of Financial Innovations and Technologies (PROFIT) with a risk-sharing facility of US$20 million. PROFIT is aimed to help extend the reach of rural financial services in the project area, in partnership with different types of financial institutions including savings and credit cooperatives (SACCOs), micro-finance institutions (MFIs) and commercial banks. The development of such financing arrangements would be eligible for support from the PROFIT innovation fund. In this context, favourable lending conditions may become contingent upon complying with sustainable agricultural practices. Another reason for moving away from a classic PES scheme was that the costs of MRV of the delivery of environmental services were considered to be too high and to be ineffective for a region on the scale of Upper Tana. Instead, the focus shifted towards providing support for livelihood improvements as a reward for the adoption of sustainable agricultural practices.

For that purpose, matching grants are considered to be the most appropriate tools because beneficiaries are expected to contribute cash, labour and in-kind inputs of at least 10 % of total project costs. The personal financial involvement is likely to lower the risk of project/venture failure and to increase the probability that grantholders will comply with sustainable management practices. Matching grants are meant to support commercially viable investments by farmers and farm groups that have adopted SWC practices (see orange box). In other words, the positive externalities that are generated through SWC practices (e.g. improvement to the environmental service 'clean water') are remunerated through access to matching grants. It is unclear, however, whether these schemes will be complemented by GWC to reward farmers for specific SWC activities. UTaNRMP prefers to provide rewards for environmental services in the form of commercially sustainable investments in improved soil and water management, rather than cash payments; i.e. rewards should come in the form of profits rather than grants or subsidies.

[4] Since most of the population do not have savings at their disposal, it is necessary to apply for affordable credit products that allow people to invest in a business or in the improvement of their livelihood. The objective of the guarantee scheme is to help them access credit without the hassles of collateral security from the eligible institutions (commercial banks).

3.4.2.2 Collaboration with Government Institutions

The design of the UTaNRMP project is in line with Kenya's long-term development blueprint enshrined in the government's 'Vision 2030' (see Box 8). This vision also includes Kenya's:

(a) Agricultural Sector Development Strategy (ASDS), which provides a framework for progressive agricultural growth and development in the next 10 years and, in this context, considers sustainable agriculture to be key to food security and to the reduction of rural poverty;
(b) Land Policy, which is aimed at a better integration of land use and physical planning from local to national levels, and at the strengthening of natural resources dispute resolution mechanisms;
(c) Legal Notice No. 166, which requires 10 % of all agricultural land to be used for forestry;
(d) Water Act, which is aimed to enhance water resources management through community participation;
(e) Forests Act to enhance conservation of forest resources;
(f) Environmental Management and Coordination Act to enhance coordination and management of activities with significant environmental impacts; and
(g) National Climate Change Response Strategy of 2010.

UTaNRMP will therefore collaborate with committees and organizations that have been set up in the course of implementing these national laws and policies in the region. These local institutions comprise not only WRUAs but also County Environment Committees, Focal Development Area Committees (FDACs) and the Community Forest Associations (CFAs). Project implementation will be based on a coordinated approach where decision making is assigned to local engaging partners whenever possible. Among other projects and programmes, close liaison with the World Bank/Deutsche *Gesellschaft für Internationale Zusammenarbeit* (German Group for International Cooperation) (GiZ)-supported Natural Resource Management Plan (NRMP) needs to continue while that project is phased out and UTaNRMP is phased in.

BOX 9

Kenya's vision of sustainability

Over the past three decades, the population of Kenya has more than tripled resulting in increased deforestation, soil erosion, and domestic and industrial pollution. Over-exploitation and encroachment of forests has reduced forest cover to only 2% of the land area. Biomass accounts for over 95% of all energy used in the country and 89% of rural households use firewood. These practices all exert heavy pressure on the protected forest areas and farmlands. Sustainable management of natural resources is a serious concern to the Government and many Kenyans. Vision 2030 aims to transform Kenya into a middle-income country 'providing a high quality of life to all its citizens in a clean and secure environment'. The strategy emphasizes the long-term development of agriculture, tourism, manufacturing and the energy sector, all of which rely heavily on sustainable management of natural resources, especially those that are associated with the five major water towers.

3.4.3 The Water Fund of the Nature Conservancy

The Nature Conservancy (TNC) was established in 1951 in the United States to protect some of the biologically important places around the world. It operates in all 50 US States and more than 34 countries worldwide and has more than 1.2 million members. The focus is on helping local communities, governments and organizations to better manage shared environmental resources. One tool TNC has developed to pay for environmental services and thus protect natural capital, is the so-called 'water fund'. Water funds engage large public and private urban water consumers involved in agriculture, industry and civil society to invest in a fund that pays for water quality and quantity upstream (see Box 9). TNC launched its first water fund in Quito Ecuador. It went on to create seven water funds in Ecuador and Colombia that help protect more than 1.58 million hectares of native habitat, which, in turn, ensures the supply of water quality for 11 million people. Each fund has a flexible structure enabling it to respond to local opportunities, needs and laws.

Apart from developing another six water funds, with the support of heavy water users downstream in Latin America, TNC is currently adapting the model to conditions in East Africa. In this context, the organization has identified the Upper Tana Basin as an appropriate site to create a water fund that would ensure the provision of water quality and quantity benefits for all users in the upstream and downstream regions of the Basin.

BOX 10

Making use of a water fund to pay for environmental services

The inspiration for the water fund came from the fund created to implement the **New York City watershed agreement**. This is widely considered to be one of the best examples of a financially sustainable public–private partnership project to ensure the quality and quantity of water through the management of a large-scale river basin that supplies water to a large city downstream (see also www.nycwatershed.org).
Water funds are part of The Nature Conservancy (TNC)'s strategy to conserve large rivers around the world, with the aim of protecting their capacity to provide the environmental services that help to ensure the supply of abundant and clean water.
One of the most successful and well-known examples is the Fund for Water Protection, FONAG, also known as the **Quito Water Fund**. This public–private mechanism was established in 2000 with a modest investment of US$21,000 and now has capital of more than US$9 million that pays for watershed conservation projects and programmes that provide water to Quito. FONAG has a board of directors formed by representatives of its constituents. This is the highest decision-making body of the fund, and is in charge of defining policies and strategies to guide the development and fulfilment of the fund's objectives. FONAG publishes annual public reports and carries out permanent audits to guarantee the mechanism's professionalism, efficacy and transparency. This builds trust and pride among investors, beneficiaries and contributors to the fund.

So far, TNC has started a feasibility study, is building a network of partners in the field and raising funds to involve important public and private sector institutions that are also large water users.

TNC plans to collaborate with the Nairobi Water and Sewerage Company, the Kenya Electricity Generating Company (hydropower), and other large water users. The aim is to jointly quantify the services nature provides in key watersheds and to develop a sustainable financing model for management programmes that would benefit conservation and business interests. This Upper Tana-Nairobi Water Fund effort builds upon the work of the recently completed feasibility study on GWC (see Box 7) in order to apply existing data, experience and expertise to the design of 'water smart' land-use management efforts in priority watersheds.

Collaboration with government water management entities, such as the Water Resource Management Authority (WRMA), will be crucial for the development of a strong institutional structure and a transparent and sustainable financing solution that can be agreed upon by key stakeholders. In addition on-the-ground implementing partners, such as the Green Belt Movement (GBM), will help TNC to guide and maximize restoration activities in the Upper Tana watershed. Using tree-planting, GBM is mobilizing communities—especially women—to improve equity, livelihoods, security and environmental conservation. The Conservancy first worked with GBM on a project that planted more than 720,000 trees in Kenya's Mau Forest over 3 years. This collaboration is being continued in the Upper Tana and Aberdare. The Conservancy is sharing its technical expertise with GBM by training their community-based programme officers in watershed management techniques. This maximizes the benefits of tree-planting efforts by focusing on the most important areas of a watershed. Other on-the-ground implementing partners that could become involved in the water fund include relevant WRUAs and other NGOs such as The International Small Group & Tree Planting Program (TIST). TNC will however focus primarily on reducing point sediment sources in general rather than on promoting good agricultural practices in particular. After all, many off-farm activities, such as quarrying, road building, dam and building construction, might also have negative effects on the ecosystem services in the watershed, especially if they affect riparian or wetland areas.

The feasibility study and the fundraising activities for the water fund were completed in 2012. The feasibility study includes economic and scientific analyses to weigh the costs and benefits of watershed conservation activities (as compared with the status quo) and to build a business case for the water fund. Once the Upper Tana-Nairobi Water Fund project has been found to be feasible, a multistakeholder steering committee will be formed to work with TNC on designing and launching the fund in priority watersheds. Through an adaptive management approach, once launched, the water fund should help to increase the focus on and impact of existing conservation efforts in the Tana Basin, leading to measurable downstream benefits to ecosystems, communities, Nairobi's water supply and Kenyan hydropower generation.

3.4.4 Lessons Learned

Effective PES schemes require trust between service users and providers—anticipating mutual compliance with contracts and excluding questionable motives. An honest intermediate broker can help bring actors with conflicting interests together, counterbalance asymmetric power relations and eventually draw up a voluntary agreement that is acceptable to all parties. In the case of the small-scale PES of Lake Naivasha, the WWF was able to achieve this role of a trustworthy broker by hiring educated local people for its local branch that successfully worked with the WRUAs that represent the buyer and seller groups but also helped mediate conflicting interests. Both buyers and sellers could expect tangible benefits from the collaboration with the WWF and were therefore willing to join the PES scheme. However, the PES scheme itself still leaves open questions. The benefits resulting from the adoption of sustainable agricultural practices are not clearly measurable, mainly because only one quarter of the upstream farmers have so far subscribed to the scheme. Those who are part of the scheme praise the assistance by the WRUAs and confirm yield increases thanks to the adoption of SWC measures. Yet the payment itself turned out to be relatively negligible (a voucher for US$70/year). Rather than being an end in itself, the payment served as a means to an end—creating new links between poor upstream players and wealthy downstream players that result in the generation of a market for environmental goods that benefit private interests but also generate numerous positive externalities.

In the large-scale watershed projects of the Upper Tana Basin, the application of PES as a tool to remunerate farmers for the generation of positive externalities seems to be even more difficult. First, the socioeconomic and agro-ecological conditions are very diverse and thus also the interests, not only differing between buyers and sellers, but also within the buyer and seller groups. This makes it more difficult to facilitate trust between actors and to reach a compromise beyond a sub-catchment. Second, the costs of MRV of the effects of changing on-farm practices on the overall water quality of the basin are very high, and the data obtained will continue to be inaccurate despite new remote sensing technologies and geographical information systems. Third, to the use PES primarily as a vehicle to facilitate communication and exchange between upstream and downstream actors may lead local markets for environmental goods, which are financially sustainable, but only if there is active support for local entrepreneurs. But such entrepreneurs would currently face many challenges because the distances between the actors are rather large, the incentives too low and the interests too diverse.

This helps explain why the managers of existing and planned watershed projects in the region consider PES a useful tool, but not the only way to remunerate activities that generate positive externalities for society and the environment. They focus instead either on concrete sustainable livelihood projects for small-scale farmers that also help to conserve soil and water (MKEPP/UTaNRMP), or on fundraising activities (TNC water fund) to mobilize the necessary means to address the environmental challenges of the Basin as a whole—going beyond mere farming activities. The overall objective is to facilitate collective action by mobilizing resources from government and non-government entities that are directly or indirectly affected by the negative externalities generated through unsustainable practices in the Basin. Potential beneficiaries of the additional means and in-kind payments available in the region through these projects must, however, prove that they are contributing with their activities to the generation of positive externalities or to the reduction of negative externalities. This conditionality is similar to PES but it leaves more choice of activities. For poor rural households it might be more important to become beneficiaries of livelihood projects that generate positive externalities. This would also automatically reduce the negative externalities that were previously caused by unsustainable practices that resulted primarily from poverty (e.g. soil depletion, deforestation, cultivation in riparian areas and on steep slopes). If livelihood projects also result in growth-oriented businesses in the region, they may also contribute to more off-farm employment and thus to structural change in the countryside. Structural change is necessary in all three of the cases studied because of the need to move people away from agriculture in order to prevent the continuous shrinkage of farm sizes. There should in fact be a threshold beyond which the size of a farm is declared to be too small to be sustainable, in terms of food security as well as environmental management.

3.5 Positive Externalities from Waste Recycling: The Maasai Slaughterhouse

Livestock and Biogas

Keekonyokie Slaughter House in Kiserian

Project Milestones: *initiation: 1981, registration as a company 1981, Biogas plant in 2005*

Environmental Service: *Waste from the slaughterhouse converted into manure for farmers and biogas. The use and effluence of waste water is minimized. Biogas is a energy substitute for wood in poor households, it therefore results in less deforestation*

Main Stakeholders: *Keekonyokie Butchers Limited, KeekonyokieMarket Access Company (MAC), The Pastoralist Field School (PFS)*

Positive Externalities: *In addition to environmental services, the business generates positive externalities in terms of social empowerment and reduction of economic uncertainty without remuneration*

Source of Funding: *Community initiative that generates its own revenues*

Innovation: *Combining meat value chain business with maasai traditions and environmental services*

Contact: *Michael Kibue (sardlivestock06@yahoo.com)*

3.5.1 Introduction

The Keekonyokie Slaughter House was founded in 1981 by a group of Keekonyokie Maasai in the town of Kiserian, south of Nairobi without external support or funding. Its purpose was to offer better market access to Maasai pastoralists and to enable them to become part of a community-owned business that not only generates its own revenues but also assists its members with many social institutions that are designed to improve the health and education of their families as well as maintaining Maasai traditions. In 1992 the hitherto informal business was registered as a company under the name of Keekonyokie Butchers Limited. The business operates in a large and complex market system involving a terminal livestock market that interfaces with a meat market and other associated service providers.

In 2005 the company started to construct and operate a biogas plant with the support of GiZ. The purpose was to convert slaughterhouse waste into energy and bio-fertilizer. This was part of an effort to fulfil National Environmental Management Authority (NEMA) requirements for slaughterhouses to stop discharge of waste water into rivers as this was causing environmental degradation. In 2011, the company presented a prototype for the packaging of biogas, which it intends to sell to local residents in addition to its existing sales of biogas to larger local institutions and its use as an energy source for the slaughterhouse.

The case of the Keekonyokie Slaughter House illustrates how a bottom-up initiative that is based on pro-poor innovation can generate a lot of positive externalities in a rural area. However, the institutional environment needs to creates the right incentives and there has to be a supporting network of organizations that have the expertise and the financial means to effectively respond to the needs of the business and help ensure its success in business, environmental management and social support services.

Unlike in the classic PES scheme, where the initiative comes mostly from a mediating institution that brings potential buyers and sellers together to negotiate a voluntary contract, ensures compliance and funds the initial stage of the project, the initiative in Kiserian came from the local people themselves. Since 1981, Keekonyokie has been managed by the community members with a board of directors, consisting of seven members. The board is elected annually based on the year's performance (indicators: income generated and volume of business). This case study illustrates how the pre-existence of local initiative and local business leadership ultimately helped to add all kinds of new social and environmental services to the business with technical and financial support from outside organizations. This ultimately led to the financial sustainability of a rural business project that also generates significant positive externalities for society and the environment. In other words, there must be a public interest in the business's continued existence in order to preserve its positive externalities. In this particular case, however, it is not market failure that justifies the remuneration of positive externalities but the fact that a market has been created that should be rewarded for its public-good character. The managers of the slaughterhouse have plenty of ideas on how to invest a financial award from the government because the slaughterhouse can hardly cope with the demand and still needs to overcome many technical and non-technical barriers to make its new environmental products (organic manure, biogas) a commercial success.

3.5.2 *The Success Story of Keekonyokie Slaughter House*

The slaughterhouse grew from processing an average of 30 cattle per day in the 1980s to around 180 per day in 2010. It attracts pastoralists from the southern rangelands and parts of the rift valley and northern Tanzania. Every day an average of 200 cattle and 400 sheeps valued at Kenyan Shilling (KSh) 5 m (around US $60,000) change hands in the market for slaughter, resale or restocking. The slaughterhouse itself has a daily turnover of between KSh 60,000 and KSh 100,000 (US$800–1200) in slaughter fees and employs 172 people. Many more are employed in associated service businesses, such as cleaning, money transfer, meat transport and other value addition activities.

The slaughterhouse has created the critical positive enabling environment for a vibrant local private sector meat business to operate. It offers a 'guaranteed' market for livestock and various kinds of social support mechanisms to its members, therefore greatly reducing the risk associated with the sale of livestock (see Box 10).

BOX 11

Business, social security and Maasai tradition

In 1981, 16 members of a Keekonyokie pastoral community decided to establish a slaughterhouse in Kiserian. The purpose was to be able to sell their meat directly in urban markets. This helped the Maasai pastoralists get 30% more on the price for their livestock. It eventually made the Keekonyokie livestock market the largest pastoralist-owned livestock marketing and processing enterprise in East Africa. It has given positive livelihoods to more than 100,000 households in Kajiado and northern Tanzania.

It eventually organized itself as an integrated value chain where its members benefit from the business as sellers of cattle and stockholders of the slaughterhouse company. The business was not only designed to generate revenues but also to reduce risk and uncertainty for its members, to serve as a social insurance mechanism and to support Maasai cultural activities. It also provides the following benefits.

1) Reduction of risk and uncertainty: In order to ensure continuity of business, even during droughts, each member contributes between one and five cattle for the purpose of restocking. These animals are bought through the revenues of the slaughterhouse.
2) Social insurance mechanisms: Contributions to health, education, and funeral expenses for members and their families.
3) Support for Maasai cultural activities: Members meet once a month to discuss the appropriate share of the revenue for each member minus social and cultural contributions. For special cases and occasions individual members contribute extra funds.

Keekonyokie Slaughter House provides a slaughter service to livestock traders, meat brokers and butchers. It is an integrated business incorporating meat processing, a meat wholesale 'supermarket', cold storage, hides and skins and intestinal meats that are popular with the bottom market segment. Other value addition activities have been introduced recently which include meat deboning, mincing and vacuum packing.

In 2005, the slaughterhouse gained some additional technical and financial support from the GiZ when it decided to invest its resources in the construction of a fixed dome biogas plant to convert the liquid animal waste into useful new products. The biogas plant has a capacity of 450 m^3. It can store 200 m^3 of biogas that is piped to a generator with a capacity of 20 kVA (Kilo Volt Amperes). Currently the biogas is being used to generate electrical power for the meat cold room, meat processing equipment and hot water for sterilizing and washing the abattoir. Excess biogas is also sold to large local institutions, such as hotels, that come and pick it up themselves. Various prototypes have already been developed to supply biogas to the local communities as cheap cooking gas to replace the charcoal obtained from cutting down trees. It started with the idea of filling discarded car tyres with biogas and then selling the sealed product to local households. However, the government did not approve of such products for safety reasons. The new product prototype consists of a refillable gas container (see Fig. 3.2). This business idea has been submitted to the Africa Enterprise Challenge Fund (AECF), which provides grants and interest free loans to businesses wishing to implement innovative, commercially viable, high-impact projects in Africa.[5] In addition the plant produces 10 Mt of liquid fertilizer per day. The liquid organic fertilizer is sold to local farmers to replace the expensive chemical fertilizers and increase the food self-sufficiency of local communities.

It was, however, not just the promising business idea that led to this investment but also the need to comply with new NEMA requirements for slaughterhouses to stop the discharge of waste water into rivers. In other words, there was a stimulus for environmental innovation thanks to stricter national environmental regulations, and the entrepreneurial board of directors of the slaughterhouse business responded actively rather than reactively. This sort of corporate leadership may be crucial not just in business but also in PES projects that aim to become financially sustainable.

3.5.3 Keekonyokie Market Access Company

The slaughterhouse business also benefited from the services of a market access company (MAC), which is part of the Rural Knowledge Network (RKN) Concept developed by FAO with the purpose of making better use of information and communication technologies to facilitate market access for small-scale farmers and pastoralists. It links value-chain players to markets, finance and market information (Nyende 2011; FAO 2011c).

[5] http://www.aecfafrica.org/

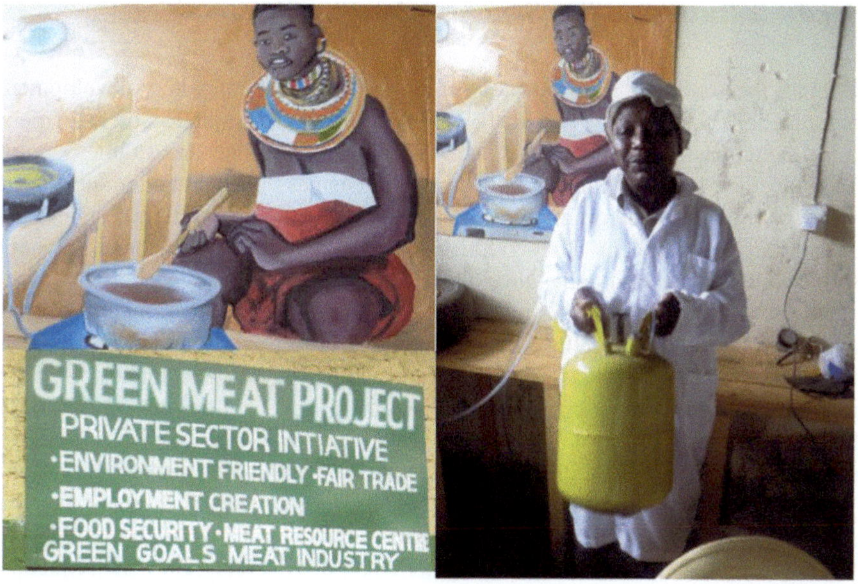

Fig. 3.2 Pictures of two product prototypes for biogas containers: a used tyre and a transportable metal container (pictures made by the author)

The MAC is currently piloting the Transaction Security Services (TSS). This system provides the buyer and seller with the delivery of the agreed volume and quality of produce at an agreed time and place. TSS is a branded service with a well laid out process and protocols for its operation and quality assurance. In the case of the slaughterhouse business it involves buying livestock from producers and tracking all the intermediate costs to arrive at the net profit. This is then shared transparently as a bonus. The TSS model therefore promotes traceability and transparency along the chain. MAC is also providing a service to the informal meat processors by charging a fee for vacuum packing and meat mincing. Finally the MAC is also the force behind the Keekonyokie Pastoralist Field School. The school was established to serve as a local learning centre through which pastoralists engage in and learn various livestock production technologies appropriate to their own value systems. It uses visual aids to illustrate various aspects of livestock production, marketing, natural resource management, and policy analysis. It also has demonstration sites for pasture conservation and water harvesting.[6]

[6] See also http://www.fao.org/fileadmin/user_upload/pes-project/docs/FAO_RPE-PES_Kiserian_Kenya.pdf

3.5.4 A Business That Generates Positive Externalities

Overall, the slaughterhouse business of Kiserian demonstrates how development organizations that seek to improve livelihoods and the natural resources of poor farmers and pastoralists can build on existing local initiatives and enable them to add value and sustainability to their particular business. The slaughterhouse business was initiated as a community initiative by a Maasai group in the early 1980s without any outside help. They gradually learned how to run their business more successfully. This eventually gave them the confidence to register as a formal company in the early 1990s. On the one hand, doing this allowed them to attract more investment while, on the other hand, they had to learn how to comply with the formal law of Kenya. This was one reason why they started to make long-term investments such as in upgrading the slaughterhouse facilities to comply with hygiene standards, and installing a biodigester to reduce waste water, to generate energy from biogas and to produce manure for the local farmers.

Figure 3.3 shows how the business functions: Pastoralists (1), who are also the shareholders of the slaughterhouse business, sell their cattle to traders at the livestock market in front of the slaughterhouse in Kiserian. They subsequently pay the slaughterhouse (2) for processing the meat, which they then sell to brokers who come and pick up the packed meat to sell to the wholesalers and retailers who supply the market for meat processing and the consumer market in Nairobi (3). The market access company (4) ensures that pastoralists are not short-changed in this

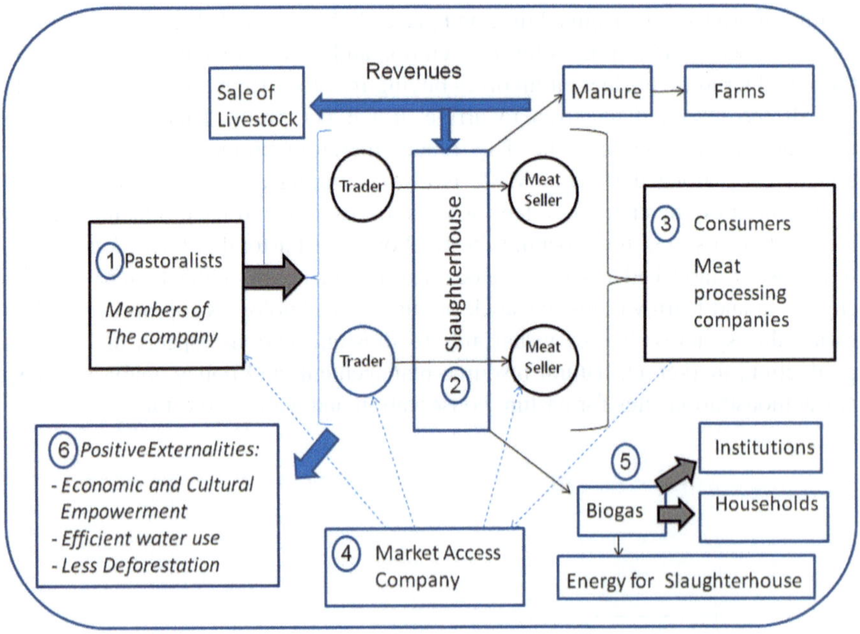

Fig. 3.3 The actors and functions in the Keekonyokie Slaughter House business model

value chain because of their ignorance about demand and supply. Transparency and traceability improves their bargaining power but also increases their incentives to be honest about the value of the cattle they sell. Finally, there are the new by-products that add value and sustainability to the business. Thanks to the construction of the biodigester the slaughterhouse is able to produce biogas for the local energy market and organic fertilizer for farmers. These activities help to save water and they decrease the dependence on charcoal as the main source of energy in the region, which helps to decrease the need for deforestation. In addition, the slaughterhouse business has become a very important employer in the region and uses every single part of the animals (including bones, hides and horns), which limits waste.

However, if one visits Kiserian, it can still look frightening and messy because the slaughterhouse occupies a large space in the centre and the small town is barely able to cope with all the cattle traders and the waste produced. But it must have looked even worse 20 years ago, when there was hardly any technology available to cope with the waste. The Maasai board of directors expresses pride in what they have achieved for their business, their community and their families without having to hand the control of their business over to outside owners. At the same time, they have reached a point where they are not asking for donations but for investments, not just because of the financial contribution but also because of the know-how they can expect to access through a professional investor. In the meantime, the slaughterhouse business produces many positive externalities for the Maasai pastoralists, who are not simply generating additional revenues through greater market transparency and shares in the companies but have also become active learners themselves through attending trade fairs and pastoralist field schools. In addition, the business may slowly move from being a generator of negative externalities to become a generator of positive externalities for the environment thanks to its investments in compliance with hygiene standards and the biogas digester that allows the business to sell two additional products with public-good character (manure for replenishment of the soil of nearby farms and biogas energy for households and hotels, which replaces charcoal and thus contributes to the reduction of deforestation). This case is crucial to illustrate that the RPE must not only be about addressing market failure but it should also reward market success that proves to be economically, socially and environmentally sustainable.

3.6 Microinsurance and Sustainable Agriculture: The Case of Kilimo Salama

Microinsurance & Sustainable Farming
The case of Kilimo Salama

Project Milestones: *Start in 2008, pilot project in 2009, 70'000 farmers insured in 2012*

Environmental Service: *improved farmer response to climate change adaptation and mitigation, increased willingness of farmers to adopt sustainable agricultural practices*

Main Stakeholders: *Safari Com, UAP Insurance, Syngenta Foundation*

Positive Externalities: *Reduction of risk and uncertainty in small-scale farming, improved access to essential farm inputs, productivity increases*

Source of Funding: *Syngenta Foundation, Global Index Insurance Facility, IFC*

Innovation: *making use of modern information and communication technology to minimize transaction costs and make insurance of agricultural inputs accessible to small-scale farmers*

Contact: *Rose Goslinga (rose.goslinga@syngenta.com)*

3.6.1 Introduction

Around the world, millions of smallholders are exposed to increasingly volatile weather conditions as a result of climate change. The agricultural sector in Kenya employs over 75 % of the country's workforce and accounts for approximately 51 % of Kenya's GDP. Yet with the exception of the export-oriented horticulture, floriculture and dairy businesses, Kenya's agricultural sector is characterized by low yields and little investment in irrigation, roads, agricultural inputs, extension, and marketing. The lack of investment in domestic agriculture is not surprising considering that the average farm size in Kenya is less than 0.5 ha whereas the mean household consists of 5.1 people. In poor rural areas, farm sizes tend to be even smaller and households larger. These conditions hamper structural change and are hardly likely to allow a farm household to produce agricultural surpluses that could then be reinvested in improved input and better practices. The first step needed to change the situation would be to enable farmers to cope better with environmental variation. Droughts and floods have a particularly devastating impact on crop harvests, food availability, household debt and incentives to invest in sustainable intensification. In the face of this exposure to environmental variation and the limited means available to protect themselves from crop failure, small-scale farmers become highly risk averse and less open to the adoption of new agricultural practices even if they may eventually prove to be more sustainable. The project Kilimo Salama ('safe farming' in Swahili) has the potential to change the situation for the better by making use of modern information and communication technologies that allow farmers to insure their purchased input in a simple, affordable and non-bureaucratic way.

Kilimo Salama was launched in 2008 and initiated its pilot project with 200 farmers in 2009. At that time, Kenya was facing a serious drought, which resulted in payouts of 30–80 % of the insured value. This led to a massive increase in farmers' interest in participating in the insurance scheme in the following year and the number of subscribers increased to 12,000. In 2011, Kilimo Salama plus was launched, which covers more crops and enables farmers to insure the value of their expected harvest. In addition, a helpline was set up that not only informs farmers about the insurance scheme, but also provides extension services.

In 2013, the project comprised 42 automated weather stations that can potentially reach the area of 330,000 farmers. In 2012, more than 70,000 farmers had already purchased input insurance and Kilimo Salama launched new initiatives to start a similar project in Rwanda and to cover not only abiotic stress incidents (droughts and floods) but also biotic ones (pest and disease infestation). Calculations of the probability of crop failure as a result of a drought or flood (based on the data gathered by the local meteorological stations) are fully automated and no visits to the affected farm sites are required. Insurance of biotic stress factors, however, would require farm visits. More intensive collaboration with other regional stakeholders that offer extension services to farmers would therefore be necessary. Such stakeholders include the local offices of the MoA, the WRUAs, as well as extension workers employed by agricultural cooperatives and microloan institutions. This collaboration will also make it possible to make use of Kilimo Salama to create a market for environmental goods. For example, Kilimo Salama could add incentives for farmers to adopt sustainable agricultural practices and agroforestry in return for discounts on their insurance premium.

Widespread adoption of sustainable agricultural practices would create additional employment through businesses that specialize in the delivery of goods and services that ensure compliance with sustainable practices (e.g. selling seedlings of trees and conservation plants or building wetlands for farm communities). Apart from the positive impact on ecosystem services, the scheme would also facilitate sustainable structural change in poor rural areas as well as social empowerment.

3.6.2 *The Stakeholders Involved*

In 2008, the Syngenta Foundation for Sustainable Agriculture was looking for new ways to help small-scale farmers in Kenya to cope with the environmental variation that was expected to increase with climate change. In this context, it discovered two automated meteorological stations that were set up by the Centre for Development and Environment (CDE) of the University of Bern, Switzerland, for research purposes in the region of Mount Kenya. It realized that these meteorological stations might have the potential to lower transaction costs for index-based insurance of agriculture and thus make it profitable for an insurance company to invest and for small-scale farmers to buy affordable insurance for their input. In its pilot project, conducted in 2009 with 200 farmers, the Syngenta Foundation was able to count on two main partners:

(a) Safaricom, the leading integrated communications company in Africa, made the business transactions required to purchase insurance possible through its M-PESA and M-SHAWRI systems.[7]
(b) UAP Insurance was prepared to insure the input farmers purchased in the Ag-vet shop if they paid an insurance premium of 5 % for the input. The automated weather station then calculated the probability of crop failure. This pilot project was a big success, not least because a serious drought occurred in 2009 and most of the insured farmers benefited immediately from having purchased an insurance policy. The farmers that had bought insurance and were calculated to be seriously affected by drought or flood (based on the output of the automated weather stations) at the end of the crop period, received a text message (SMS) that announced their eligibility for compensation for crop failure and confirmed that the corresponding money transfer had been made to their mobile account. The insurance scheme became very popular because of its reliability and affordability. Moreover, farmers were encouraged by other local stakeholders to participate in the scheme. For example, two thirds of the subscribers were clients of the microfinance institutions (MFIs) that have made enrollment in Kilimo Salama mandatory.

3.6.3 Kilimo Salama Plus

In 2011, Kilimo Salama plus was introduced. Whereas the initial scheme was quite limited and was linked to Syngenta products, the new scheme offered farmers free choice. Moreover it allows farmers to cover their investment in farm inputs *and* output value. The value of the harvest is an estimate of the expected harvest value in Kenyan shillings. The crops covered by the scheme are maize, beans, sorghum and wheat. The risks covered include losses due to drought during the establishment, vegetative, flowering and silk periods of plant growth. The risk of excess rain is covered during the period of yield formation and ripening.

[7] M-SHWARI is the revolutionary new banking product for M-PESA customers that allows people to save and to borrow money through their phones while earning interest on money saved. With M-SHWARI, people are also entitled to affordable emergency loans.

Subscribers also have access to personal farming advice through a free helpline via SMS.

The insurance can be bought at selected agro-vets shops and farmers are free to buy insurance for their preferred brand of farm inputs. The purchase of the insurance is a formal business transaction that requires the farmer to be registered either by means of an ID number or a SIM card. The cost of the insurance premium is usually 10 % of the chosen value but can vary depending on the weather stations. Purchasing such input products as a farmer group allows farmers to obtain discounts on transport cost and product price.

3.6.4 Why Does It Work to the Benefit of Small-Scale Farmers?

A farmer can try out insurance for as little as ten US cents to insure a bag of seed costing two US dollars. If there is a drought, for example, the farmer will receive a payout of two dollars, and can begin afresh in the next growing season. Previously, few farmers could afford such cover because of the high transaction costs. Traditional crop insurance relies on expensive farm visits to verify claims. Kilimo Salama does not visit the farms. The programme is designed specifically for smallholders and uses automated weather stations and mobile payments. These features dramatically reduce administrative costs, enabling insurance to be bought at a price that millions of farmers can afford (see also Ferroni and Castle 2011).

Kilimo Salama's use of technology is the key to the microinsurance product's affordability and the model's scalability. As already mentioned, Kilimo Salama's clients are smallholders. To reach them, the programme partners with agricultural microcredit institutions and local agro-vets or stockists who sell farm inputs such as seeds, fertilizer, and crop protection products. When a farmer purchases insurance, the microcredit officer or agro-vet registers the purchase by scanning in a code using a specially developed mobile phone application. The message goes to a cloud-based server that administers the policies. It then sends the farmer an automated SMS with his or her policy number. Solar-powered weather stations collect the weather data. At the end of each growing season, the data collected is automatically compared to an index based on historical weather data. If the season's rainfall is 15 % above or below the average, the insurance payout owed to client farmers is calculated and sent via automated mobile payment. There is no 'claims' process.

With no field surveys, no paperwork and no middlemen, transaction costs are minimal. The scheme is therefore financially sustainable. Kilimo Salama is the first insurance scheme in the world to use a mobile network-based platform and on-the-ground solar weather stations to provide smallholders with low-cost insurance policies.

As a mark of its success, it won the *Financial Times*' award for Technology in Sustainable Finance in 2012, in recognition of the ground-breaking work done to provide smallholders with access to insurance cover using innovative technology and approaches.

3.6.5 The Link to Positive Externalities and Markets for Environmental Goods

According to Rose Goslinga (personal communication), the coordinator of Kilimo Salama in Kenya, the insurance scheme was meant to be linked to incentives for farmers to adopt conservation agriculture. Syngenta, which already offers insured farmers a 5 % discount on their premium for the purchase of its products, also sells the herbicide that makes no-tilling possible. However, the monitoring turned out to be a problem at that time because the project was understaffed. Today, Kilimo Salama works closely with the local offices of the MoA, WRUAs, and microloan institutions. Since the technical officers and extension workers of these institutions make frequent farm visits and have an interest in encouraging farmers to adopt sustainable agricultural practices, Kilimo Salama may soon be able to offer a discount on the insurance premium for farmers that are reported to have adopted such practices. Conservation agriculture could be one such practice.

Kilimo Salama has the big advantage of being run as a local business. Because of the low margins, the business depends on having a large number of subscribers to make it profitable. Since about 10 % of them are large-scale farmers, it is possible to cross-subsidize. At any rate, microinsurance has turned out to be a product that small-scale farmers really want and it has the potential to introduce many new insurance- and non-insurance-related products and services. Many of the non-insured products and services could be provided by local actors and linked to markets for environmental goods. WRUAs, for example, could make privileged access to insurance premium discounts conditional upon the adoption of sustainable farming practices and the adoption of agroforestry. It could also start its own tree nursery business or encourage farmers to invest in it. Many farmers that have gained knowledge and experience on how to convert a non-sustainable farm into a sustainable one could help other farmers comply with local sustainability standards and thus enable them to gain access to input discounts and output premium prices.

3.7 Encouraging Sustainable Flower Production Through Private Standards

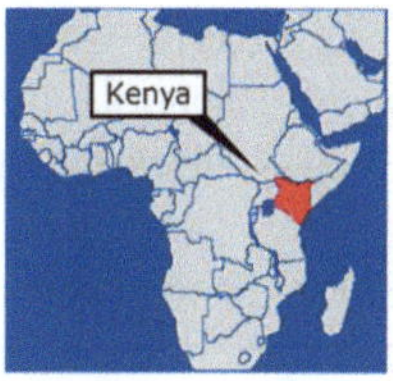

Markets for environmental goods

The case of the Kenyan flower business

Project milestones: *Creation of the Kenyan Flower Council in 1996, Water Act in 2002*

Environmental service: *Sustainability labels by third-party certification agencies in return for complying with private standards and environmental regulation > private investment in green innovation*

Main stakeholders: *Kenyan Flower Council, GLOBALG.A.P., MPS, ETI, flower business with 70,000 directly employed workers, wetland constructors, companies selling advanced greenhouse technology , University of Eldoret*

Remuneration of positive externalities: *Minimizing the environmental impact of flower production, increasing worker welfare in return for recognition labels*

Source of funding: *private sector investment*

Contact: *Jaime Hernandez (jaime.hernandez@syngenta.com, Peter Ochami, Manager of Zena Rose Kenya, Thika; Samuel Kamau, Manager of Kariki, Ltd, Thika*

3.7.1 Introduction

The floriculture industry has grown into the third-largest foreign exchange earner in Kenya after tea production and tourism. The growth of the industry has been accompanied by increasing criticism of the possible cost of this growth to workers' health and the environment (Madelev 2009). In particular, flower producers in the region around Lake Naivasha have been accused of not respecting labour rights and for causing a drop in the lake's water level, polluting the lake and possibly affecting the lake region's rich fauna and flora (BBC 2003).

In the course of the first decade of the twenty-first century the industry went through substantial changes in response to tighter public environmental regulation, water pricing and increasingly demanding private standards set by global retailers that go far beyond technical issues related to workers' health and the use of chemicals. By creating the Kenyan Flower Council (KFC) in 1996, flower growers responded quickly to the growing need to require flower export farms to take appropriate measures to improve workers' welfare and the quality of the natural environment. KFC also ensures that flower growers comply with the private business-to-business (B2B) standards set by globally active third-party certification agencies such as MPS ('More Profitable Sustainability', the Dutch standards for flower imports) and GlobalG.A.P. that act as de-facto gatekeepers controlling access to the European retail markets. Generally, the proactive stance of the Kenyan flower industry has indeed improved its environmental practices (e.g. less nutrient effluents, less pesticides, less water used) and social provisions (e.g. regular working hours, higher salaries, social safety measures, decent housing, schools for workers' families) and overall sustainability record (Riisgaard 2007; Mekonnen et al. 2012).

In response to the requirements for compliance with private standards, the flower business has invested heavily in technical skills, system-oriented agriculture and logistics. In addition, the industry is making increased use of technologies that save water, energy and chemical input and help to clean up effluents before they are released into the environment. In other words, the concern for the industry's reputation and the need to comply with private standards has once again created a market for environmental goods. One such market has emerged from the need to establish 'farm wetlands' that ensure minimal water pollution. The demand for wetlands has created an industry that is dedicated to setting up private wetlands near greenhouses. The consultants who offer to install such wetlands on farms,[8] have been a response to earlier environmental regulation in the United States and Europe. They are now expanding in many emerging developing countries (see also Sect. 1.2.3).

Furthermore, the requirement to phase out harmful chemicals has led to the development of more benign agents for plant protection and to innovative pest management strategies that reduce the number of applications needed and increase the safe use of pesticides. One problem is that all these inputs to improve environmental sustainability and ensure compliance with private standards make the flower business very capital-intensive. Small-scale producers who cannot afford such input packages, and who have not been trained in integrated pest management, are therefore increasingly considered to be a business risk because they are unlikely to be able to comply with strict private standards. As a consequence, outgrower schemes enabling small-scale farmers to participate in the flower export business have become increasingly rare. Instead these farmers may obtain their input from secondary markets that are handled by local agro-dealers. In the absence of training in pest management, means of plant protection are often handled in an unsustainable and unregulated way (Clapp and Fuchs 2009).

Even though no one would be likely to deny that the environmental impact of such a fast-growing industry as the Kenyan flower export industry is still significant, this knowledge-intensive business also has a great potential to help improve the sustainability, quality and productivity of domestic agricultural activities that are less subject to standards and regulation, harder to monitor and are thus also more prone to causing environmental problems. Around Lake Naivasha, the sources of the most serious water pollutants, according to the local WWF staff,

[8] see http://www.environmental-expert.com/companies/keyword-constructed-wetlands-3343

can often be traced back to local small-scale farmers upstream that are detached from formal markets but nevertheless buy input from the local agro-dealer to increase yields on their small plots (see Sect. 3.3).

A PES scheme should take into account that in many poor rural regions in developing countries it is poverty rather than affluence that causes most of the sustainability problems (Beckerman 2002). Consequently, upstream farmers are not only potential providers of an environmental service but are often also a significant source of environmental pollution.

In the region of Lake Naivasha, this makes it all the more important that the flower business downstream not only acts as a buyer of environmental services from upstream domestic farmers in PES schemes, but also assumes a more active role in ensuring that upstream farmers receive proper and effective advice and training in the use of legally approved means of plant protection. Some companies, such as Syngenta with its microinsurance scheme (see Sect. 3.6), are already making use of information and communication technologies to better inform agro-dealers about the sustainable use of the company's products and to organize training courses on the sustainable use of pesticides. In return, farmers have more possibilities to report overuse of plant protection products by neighbouring farmers or abuse by a local agro-dealer or sales agent who conceals important information about the sustainable use of the various input products they are selling to farmers. Agro-dealer shops also tend to evolve slowly into market points where farmers not only shop but can also test new products and techniques and attend training courses on sustainable pest management (Juma 2011).

3.7.2 The Evolution of the Flower Business in Kenya

The history of the export of fresh horticultural produce from Kenya started when the country was still a British colony. After independence, exports started to grow continuously and eventually the business became Kenya's third-largest source of foreign exchange.

Kenya's flower industry has experienced rapid annual growth rates of more than 20 % over the past 20 years. It has grown into a US$500 million export business and employs more than 70,000 workers directly and more than 1.5 million indirectly. Kenya has become the lead exporter of cut flowers to the EU with a market share of over 35 %. The largest share of exported cut flowers are roses, followed by other high-value flowers. Approximately 65 % of exported flowers are sold through the Dutch auctions, but direct sales to global retailers are growing.[9]

The industry continues to attract investors as a result of having a solid infrastructure, favourable climate, a proactive work force, and the positioning of Kenya in the global flower market. The Kenya flower industry comprises mostly large- and

[9] See http://www.kenyaflowercouncil.org/floricultureinkenya.php

medium-scale producers who comply with the high social and environmental standards set by the KFC, which ensures that the quality of the exported flowers meets the private standards set by the retailers in Europe and elsewhere. The growers have a vast knowledge complemented by modern technology for precision farming as well as prowess in marketing.

The rapid growth rates have brought the industry under increasing scrutiny by journalists and NGO activists. The flower producers have mainly been criticized for their working conditions and the impact of intensive use of crop protection chemicals on Kenya's environment and workers' health. The seasonality of the flower business, with peak demands on Valentine's Day, Mother's day and at Easter, has intensified the need for a flexible workforce, especially because foreign retailers often adjust their orders on the day of delivery. Moreover, the perishability of the product often requires workers to work long hours to ensure timely harvesting and spraying.

Kenyan producers had already designed and adopted codes of conduct in the mid-1990s to fulfil the requirements of overseas customers, mostly represented by large retailers in the Netherlands and Britain. Since then, social and environmental standards have proliferated but, unlike in the 1990s when self-regulation mostly covered the use of chemicals and environmental management, the private standards that have been established by national flower importers go far beyond technical issues. These private standards comprise MPS in the Netherlands, FLP (Flower Label Programme) in Germany, ETI (Ethical Trade Initiative) in the UK, and large European retailers (GlobalG.A.P.). They have become very prescriptive and burdensome to implement. Often they are no longer linked to scientific evidence but merely reflect presumed consumer perceptions in affluent countries. This makes the business unnecessarily capital-intensive, with the consequence that small producers that previously flourished thanks to the less strict and less buyer-driven auction system find it increasingly difficult to stay in business, and large flower producers have become reluctant to maintain the outgrower schemes that allowed small-scale farmers to participate, because they cannot fully control their practices.

3.7.3 Buyer-Driven Private Standards: Are They Still Benefiting the Poor and the Environment?

A certificate of compliance with a private standard usually represents a written assurance that a product, process or service is in conformity with the applicable standards. As such its label or symbol indicates that compliance with standards has been verified and approved by the standard-setting body that accredits certification bodies (Hatanaka et al. 2005). While the certificate is a form of communication between seller and buyer, the consumer label can also be a form of marketing strategy to attract more end-consumers.

In the Third Party Certification (TPC) system, suppliers are not dealing directly with the large retailers but with local accredited certification bodies. These national or regional certification bodies must in turn be accredited by a global standard-setting body from which they receive their accreditation. Suppliers thus receive their certification from the local branch, in this case an 'independent' TPC. But this requires that they pay the certification fee, hire consultants to help them upgrade their equipment, submit to annual inspections, conduct external audits and even pay for the analysis of residues and tests of soil core samples in laboratories (Hatanaka et al. 2005; Freidberg 2007). This may be manageable for capital-intensive, large-scale suppliers or for farmers that receive generous support from government, but certainly not for small-scale producers in developing countries—even if they are organized into cooperatives.

3.7.3.1 GLOBALG.A.P.: A Third-Party Certification Body

GLOBALG.A.P. is a not-for-profit organization that aims at ensuring safe, sustainable agricultural production worldwide through voluntary standards for the certification of agricultural products around the globe. G.A.P. stands for good agricultural practice—and GLOBALG.A.P. is the worldwide standard that assures it. GLOBALG.A.P. was originally formed in 1997 as EUREPG.A.P., an initiative taken by European retailers to respond to numerous food safety scandals and the introduction of genetically modified (GM) food in Europe. It eventually set the global private standards for product safety, environmental sustainability and worker safety and welfare. The goal was to harmonize previous private standards and procedures and develop an independent certification system for Good Agricultural Practices (G.A.P.).

The GLOBALG.A.P. system today relies on control points and compliance criteria (CPCC) not only for fresh fruits, flowers and vegetables, but also for feed crops, livestock and fish products. It encompasses 16 standards that cover more than 400 certified products. The CPCC protocol is compiled from numerous control points that take the form of 'major musts' and 'minor musts'. The CPCC themselves have been drawn up for each of these standards by a technical committee of representatives of retailers and suppliers. The benchmarking process is based on a one-to-one comparison principle where private or public schemes existing in different regions or countries are contrasted with GLOBALG.A.P. These schemes usually address certain requirements identified for the particular geographical locations and marketplaces, which need to be met to qualify for the GLOBALG.A.P. standard. Countries that are not yet in a position to meet the high global standard can first become eligible for the LocalG.A.P. standard. This provides an achievable level of assurance and reliability that helps producers gain access to new local and regional markets. There is also a GLOBALG.A.P. add-on, which allows producers to be recognized for going beyond the global standard. This raises their status as producers and offers buyers specific assurances tailored to their interests and preferences.[10]

[10] http://www.globalgap.org/uk_en/index.html

GLOBALG.A.P. recognizes other farm assurance schemes via benchmarking. Accredited certifying bodies on the national and regional scale must follow strict criteria to become eligible. Once they comply with all requirements they become part of the online certification database that customers can use to check up on producers and validate certificates. To ensure that accredited certification bodies conduct their audits in line with GLOBALG.A.P. requirements and procedures, a Certification Integrity Program (CIRPO) has been put in place that issues warnings, or even removes those organizations that repeatedly fail to comply from the database. Currently there are 1400 trained inspectors and auditors working for 142 accredited certification bodies to perform independent third-party producer audits and issue certificates for 130,000 producers in more than 110 countries.

In developing countries, national G.A.P. schemes (as entry-steps to GlobalG.A.P.) have been established with the support of governments. They have almost become a prerequisite for enlarged market accessibility, including for small-scale farmers. Small-scale farmers would not be able to comply with the G.A.P. private standards requirements (especially if they are illiterate) unless they are assisted by consultants from governments and NGOs. These consultants often do not work with single farmers but with farmers' cooperatives in order to obtain group rather than individual certificates, as these are cheaper and require less individual effort. Nevertheless, even under these circumstances, many retailers are reluctant to source from suppliers with outgrower schemes because ensuring that outgrowers are following every little step correctly is nearly impossible.

This is especially true for agricultural products where GLOBALG.A.P. is just one of the players in the business of global standard-setting. As an international standard-setting body it is in competition with many other B2B labels, such as the Dutch MPS-ABC scheme in the flower business which aims at phasing out almost all means of plant protection classified in toxicity class 1. This makes hardly any sense because, as has been well known since Paracelsus that it is the dose makes the poison. Often the rigid approach leads instead to an increased use of other means of plant protection that may be harmless in small doses but become harmful once they are used on a wider scale.

3.7.4 B2B and B2C Labels in the Kenyan Flower Industry

In the case of Kenya's cut-flower export industry, where outgrower schemes have been nearly abandoned, it is not uncommon for a producer to hold five or more different certificates. Some may be individual company standards imposed by large retailers, which are adopted throughout their supply chains and are mainly designed to communicate a commitment to sustainability to consumers (e.g. Tesco's Nature's Choice, Carrefour's Filières Qualité, and Coop's Naturaplan); others are collective national standards set by organizations that operate within the boundaries of individual countries, including industry associations and NGOs. In Kenya, the KFC administrates one of the most prominent national standards. The KFC Code of Practice (CoP) has references to more than 23 different government statutes relevant to floriculture: the principles of the International Labour Organization (ILO), International Code of Conduct (ICC), Ethical Trade Initiative (ETI), as well as the Horticulture Ethical Business Initiative (HEBI). It is also fully benchmarked to GlobalG.A.P., Fair Flower Fair Plants (FFP) and many other international and individual standards that apply in the flower trade. With its CoP, KFC ensures that growers comply with the ever changing international and local standards.

The main thrust of the KFC CoP is social welfare, safety and health at the workplace, environmental conservation and protection, good agricultural practices, training, and quality management. Complying with the KFC CoP is no longer just about demonstrating CSR but has become de facto mandatory for the flower export industry. Generally, the proactive stance of the Kenyan flower industry has improved its record of environmental (e.g. less nutrient effluence, less pesticides, less water) and social (e.g. regular working hours, higher salaries, social safety measures, decent housing, schools for workers' families) sustainability (Riisgaard 2007; Mekonnen et al. 2012). Despite its size and growing economic importance in Kenya, the Kenyan flower industry is considered to be one of the less polluting agricultural producers thanks to substantial investment in sustainable technological change and continuous upgrading. Exporting flowers in Kenya has however become a capital-intensive business that is increasingly detached from the local agricultural economy.

3.7.5 The Private Standards Landscape in the Flower Business

The private standards 'landscape' is highly dynamic and has led to a confusing variety of different types of process-oriented standard-setting bodies. Of these countless sustainability labels, no more than a handful of standards are explicitly communicated to European consumers by means of a consumer label (B2C). The leading label in terms of flowers sold is believed to be the Fairtrade label (Fair Trade Labelling Organization (FLO)/Max Havelaar). However, the share of labelled flowers in the European consumer market has been rising quite rapidly over the past few years. Sales of Fairtrade flowers, for example, have more than tripled since 2004. In addition, an increasing number of supermarket and garden centre chains have decided to use their own private labels to communicate whether an item has been produced in a fair and sustainable way. In this context, it could be safely argued that the closer the label is to the consumer, the more likely it is that the label is less about communicating a message about product/worker safety and environmental sustainability than about communicating values as a form of reputation management and product differentiation (Henson and Humphrey 2010). This trend is obvious not only in the flower retail business but in retailing in general.

Tesco's Nature's Choice, Carrefour's Filières Qualité, Marks & Spencer's Field-to-Fork, and Coop's Naturaplan, are just a few examples. In addition, there are a wide range of private standards and labels covering whole farm sustainability, specific crops, organic products and food safety.

3.7.6 The Problem with B2C Private Labels

If B2C standards are mainly tools for communicating values as a form of reputation management and product differentiation, then questions must be raised as regards the added value in terms of environmental sustainability in agriculture, the claims of adhering to principles of openness, transparency and consensus, and the compatibility with the principle of non-discrimination. It may be true that B2C labels are mostly based on B2B labels and that there are numerous hybrid forms where the B2C label is handled by an independent private entity or a joint national retailer initiative (e.g. the Business Social Compliance Initiative (BSCI) in Switzerland). However, there is still one big difference that makes the distinction between B2C and B2B increasingly dubious: B2C labels obtain a price premium for being sustainable and fair, whereas B2B labels do not, even though they have the same social or environmental performance or sometimes even exceed B2C labels (Modelo 2014). Moreover, the higher price for B2C labels (organic, fair trade) tends to signal to the consumer that he or she can either buy the expensive but fair product or the cheaper but less 'fair' product. This is clearly misleading since the 'cheaper' version often performs better in terms of social and environmental sustainability than the expensive version (Aerni 2013).

There are other worrying trends with B2C labels that are most obvious with the 'free from' labels. Products that are marketed as being 'free from' genetically modified organisms (GMOs), aspartame, monosodium glutamate or parabens tend to be more about values and common fears than about scientific evidence. It could even be argued that by labelling products as 'free from', supermarkets are playing on people's fears, which are based on the rumours that have circulated about these substances (Murphy 2013). B2C labels are therefore more likely to contravene the WTO principle of non-discrimination than B2B labels in the food chain; even though they may often be conceived as just a cheap add-on to B2B labels and national bans that are already in place (e.g. the ban on GMOs in most European countries) and therefore do not affect international trade directly.

B2B private standards and the associated B2C private labels have also been subject to broader criticism. They have led to audit fatigue, confusion of consumers and suppliers, duplication (with the multiple overlapping audits per supplier), inefficiencies, high costs and a focus on audits rather than remediation. The Global Social Compliance Programme (GSCP) has been set up by the leading retail companies to address the need for consistency as well as the root causes of non-compliance. It will assist retailers and brand manufacturers around the world, in whatever industry, to work towards mutual recognition of audit results. This may make business easier for suppliers but it still does not address the growing confusion that consumers face when confronted with all kind of 'goodness' labels (Freidberg 2007). It may also have a negative effect on the social and environmental sustainability performance of Kenya's flower exporting business because its fear-based labels may act as a brake on environmental innovation (especially when biotechnology is involved). This is likely to further increase the barriers to market entry for small-scale farmers and fail to take into account important insights by farmers and input providers on the optimal and most sustainable use of means of plant protection.

3.7.7 Concluding Remarks

Since the turn of the millennium, the flower export industry in Kenya has faced ever tighter public environmental regulation, water pricing and increasingly demanding private standards set by global retailers to comply with social and environmental standards. Flower exporting companies had already anticipated these changes in the 1990s and created the KFC in 1996 to respond to the growing need to require flower export farms to take appropriate measures to improve workers' welfare and the quality of the natural environment. As a result, many companies invested in innovation to comply with new public and private regulation. Such innovations, including the creation of farm wetlands to minimize the harmful effluents, more benign means of plant protection, more products to ensure workers' health and safety and more extensive training on the safe and sustainable use of inputs, have become more widespread. All these efforts have contributed to the improvement of environmental services—but since these improvements happened through innovation rather than regulation, they have hardly been noticed by public officials and NGO representatives involved in projects designed to improve the provision of environmental services.

The reason for not counting green innovation in the private sector as a genuine contribution to the improvement of environmental services, has a lot to do with the general perception that the private sector would merely be concerned with securing short-term profits. But this attitude ignores the fact that the flower export producers are very concerned about their reputation with end-consumers in affluent markets. It is in their own interests to be perceived as sustainable and fair market players by showing an increasing willingness to participate in joint efforts to address local and environmental problems through financial and technical assistance. For example, they participate as the main buyer party of environmental services downstream in voluntary watershed PES projects and thus help to compensate upstream farmers for the adoption of sustainable agricultural practices. All this might be at the expense of short-term profits but it still makes business sense from a long-term perspective. It is unfortunate, however, that retailers in affluent countries still prefer to portray the sustainable food choice for consumers as a binary one (i.e. organic/

fair trade = sustainable and expensive; all other products = unsustainable but cheap).

The result is that greater investment in sustainable production by global agricultural companies does not translate into an enhanced reputation with affluent consumers. Retailers therefore increasingly put the brakes on green innovation in agriculture despite their claims to promote it in behalf of their customers.

Chapter 4
Conclusions

Abstract The chapter summarizes the insights of the previous chapter. It shows that market-based instruments in general and PES in particular only work if they encourage local entrepreneurship and innovation, if they are well-embedded into existing macro- and meso-economic institutions, if they take into account the level of economic development and the specific ownership structure. The chapter concludes by arguing that the real success of PES and other market-based institutions may be rooted in its capacity to serve as a vehicle for the creation of local markets for environmental goods that not only improve the environmental sustainability of agricultural practices but also stimulate local economic development.

Market-based instruments in environmental policy and management, such as Payments for Environmental Services (PES), are aimed to address market failure by encouraging affluent beneficiaries of environmental services and owners of ecologically valuable resources to participate in a voluntary agreement that is based on a quid pro quo bargain (payment in return for the adoption of practices that ensure the sustainable provision of environmental services). In this book we have shown a wide range of applications of such market-based schemes. These include classic third-party funded PES projects on small and large scales, incentive-based agro-environmental policies in developed and developing countries, global and national emissions trading markets designed to integrate local communities as potential sellers of environmental services, and the design of private standards for good agricultural practices designed by global retailers and commodity organizations to encourage the local producers to make sustainable use of ecosystem services in return for a better farm-gate price.

The effectiveness of all these instruments, policies and initiatives depends on the way they interact with the existing macro- and meso-economic institutions, the level of economic development and the specific ownership structure. Yet, while voluntary private standards designed by farmers and commodity organizations to improve sustainable agriculture (B2B labels) and incentives by input providers to provide discounts to farmers in return for the adoption of sustainable practices may well work, there is still no evidence that classic PES projects, carbon sequestration markets for smallholders and private standards designed by global retailers primarily to please western consumers (B2C labels) are contributing to increased sustainability and productivity in the long run (Fisher et al. 2009; Gómez-Baggethun

P. Aerni, *The Sustainable Provision of Environmental Services*, CSR, Sustainability, Ethics & Governance, DOI 10.1007/978-3-319-19345-8_4

et al. 2010; Modelo 2014). Often they tend to discourage local entrepreneurship and innovation because the terms of cooperation are set in advance by the external donor or buyer, and the standards they develop fail to take into account the value of innovation, which by definition, is not captured by standardization.

Many scholars criticize current market-based environmental policy instruments such as PES for:

a) not adequately addressing the many well-known challenges in institutional economics and political economy, such as the problem of asymmetric information and the inadequate alignment of private incentives with the overall public interest (Kinzig et al. 2011; Kosoy and Corbera 2010; Pirard 2012a; Muradian et al. 2010; Pirard et al. 2010; Ferraro 2008, among others).
b) being based on a fictitious market, considering that demand and supply have to be organized and incentivized by an external mediator (Van Hecken and Bastiaensen 2010; Vatn 2010).
c) treating environmental services like a commodity that can be bought, sold and traded. This assumption is on shaky ground considering the complexity of the underlying social, political and biophysical relationships between humans and the environment (Kremen 2005; Norgaard 2010; Swift et al. 2004).
d) moving from the 'polluter pays' principle in theory to 'pay the polluter' in practice because of the attempt to simultaneously achieve environmental sustainability and poverty reduction in agriculture in low-income countries (Hanley et al. 1998). Addressing two goals with one tool is not effective, as illustrated by Tinbergen (1952) and confirmed in the case of watershed PES in Kenya as discussed in Sects. 3.3 and 3.4.

Despite these valid criticisms, many projects designed to improve the sustainable use of natural resources through market-based institutions such as PES seem to be working well for the time being—as illustrated by this book, as well as by the factsheets contributed to an FAO multistakeholder conference organized by the author in September 2013.[1] Yet, a closer look at the examples described reveals that PES works for reasons that were not anticipated by PES theory and that have little to do with the incentive provided by the payment itself. In the context of PES watershed projects that include an upstream and a downstream region, voluntary PES schemes are often complemented by government policies that include additional sticks and carrots to ensure that the private interests of the different parties upstream and downstream are better aligned with the public interest to improve water quality to meet targets measured at the border between the two regions (Wen et al. 2012). There are also many forms of in-kind payments provided by the mediator of PES, as well as the social network created in the course of the implementation of PES projects that help create new opportunities for revenue generation by local participants (Prager et al. 2012). In other words, they are hybrid PES systems that have departed to a significant extent from the classical PES

[1] See also http://www.fao.org/nr/aboutnr/incentives-for-ecosystem-services/case-studies/en/

principles. Finally, increased access to knowledge and user-friendly technologies thanks to the introduction of a PES scheme may lead to innovative activities that are not only able to address environmental problems cost-effectively, but also result in markets for new goods and services that meet local needs and improve overall environmental management (Cohen and Winn 2007). This eventually leads to a process of social empowerment that is often driven by indigenous people who strive to generate their own revenues by creating markets for environmental goods and services and to become less dependent on external funding with strings attached.[2]

In sum, the real success of PES and other market-based institutions may be rooted in its capacity to serve as a vehicle for the creation of local markets for environmental goods that not only improve the environmental sustainability of agricultural practices but also stimulate local economic development. Even though the role of local entrepreneurship and innovation has been acknowledged in the landscape approach (Prager et al. 2012), publications on the potential of hybrid PES systems to create financially sustainable markets for environmental goods is still sparse. This is in contrast to the growth in literature concerning Conditional Cash Transfers (CCTs). This literature focuses increasingly on the need to make the achievements in poverty reduction and enhanced education enduring through more investment in the creation of knowledge-intensive off-farm employment (Rawlings and Rubio 2005).

Whether incentive-based schemes to improve environmental services do indeed achieve their envisaged environmental goals, i.e., to enhance and sustain the provision of environmental services, may ultimately depend on their capacity to generate local revenues. This provides additional incentives for local participants to become actively involved as entrepreneurs.

But if that is the main insight, PES and other market-based institutions to improve environmental services should reframe the whole challenge. In future, it should be less about addressing 'market failure' through regulation and fictitious markets that depend heavily on external funds, and more about facilitating the creation of new markets for environmental goods through public investment in local entrepreneurship and innovation, and private–public partnerships that enable such markets to grow and thrive.

These lessons can be learned from the experience in the South. Kenya's policies and its projects designed to improve environmental services are increasingly based on experimental rather than social planning approaches. It is trial and error and learning by doing that often help resolve the apparent trade-offs between protecting the environment and improving the economic situation of the poor. It is their capacity to innovate, or merely to adopt innovation, that matters and needs to be supported by the government, academia and the private sector.

Western scholars often fail to understand why people in the South find that their well-designed market-based institutions, derived from the principles of neoclassical

[2] See also the findings of the most recent conference on PES in Costa Rica, co-organized by FAO: http://www.fao.org/docrep/019/i3754e/i3754e.pdf

economics, are not very helpful in efforts to improve the quality of ecosystem services while also improving the economic situation. PES scholars must learn that the root of the sustainability problem in developing countries is poverty, not affluence. Poverty was also the major cause of the sustainability challenges in Europe in the nineteenth century (especially the massive deforestation and the increase in the problems of waste disposal as a result of population growth). But at that time, policy makers in Europe would have equally rejected policy instruments that framed technological and economic change as part of the problem rather than part of the solution—because people needed to be lifted out of poverty through education and integration into the formal economy. Governments in Europe back then did not have the budget to subsidize the protection of environmental services but instead had to rely on public–private partnerships to cope with the growing sustainability challenges through innovation in technology and management (Aerni 2007, 2009). In other words, governments had to be facilitators, not just regulators of change. The same is true today in many developing countries, they have neither the time nor the financial endowment to search for the perfect solution or to continue with approaches that look nice in theory but do not work in practice. They rely on trial and error as Europe did in the nineteenth century. But, unfortunately, many people in affluent developed countries are not very interested in the question of how their own societies had to embrace change to achieve a level of economic development at which an increasing share of the surpluses and taxes could be invested in the improvement of societal and environmental conditions. Instead they wonder why economic and technological change was necessary in the first place. While taking it for granted in their daily lives and lifestyles, they increasingly resent the risk and uncertainty that is inherent to change through innovation. Joseph Schumpeter highlighted early on how this public attitude in affluent societies eventually leads to increasing bureaucratization and the decline of entrepreneurship-driven and disruptive capitalism (Schumpeter 2008).

Orthodox environmental economists that follow the principles of neoclassical welfare economics with its comparative-static and ahistorical baseline assumptions tend to ignore the fact that different stages of economic development require different regulating and facilitating policies to generate desirable societal and environmental outcomes. More interdisciplinary social science research is therefore necessary to bring together different social science disciplines that are generally not based on the shared implicit baseline assumptions but challenge them.

Annex: Recorded Interviews in Kenya (August 2012)

KACP Project

Henrik Brundin, Regional Director of Vi-Agroforestry, Nairobi
Bo Lager, Programme Director of Vi-Agroforestry, Nairobi
Wilson Odongo Nyariwo, Project Manager of Vi-Agroforestry, Kisumu, Nyanza
Caroline Musee, Monitoring and Evaluation Officer at Vi-Agroforestry, Kisumu, Nyanza
Morris Karadha, Farmer involved in the KACP Project, Kombewa, Nyanza
Francis Olenyo, Farmer involved in the KACP Project, Kombewa, Nyanza

Lake Naivasha Small-Scale Watershed PES Project

Robert Ndetei, WWF, Project Coordinator, River Malewa Conservation Project, Naivasha
Peter Muigai, WWF, Technical Officer, River Malewa Conservation Project, Naivasha
Josephat Nyongesa Mukele Josephat, WWF, Natural Resource Economist, Naivasha
Enock Kiminta, Lake Naivasha Water Resource Users Association, LANAWRUA, Naivasha
James Waweru, General Manager of the Flower Business Park, Naivasha
James Wainaina, Farmer involved in the PES Project, Engineering Town
Margret, Widow farmer involved in the PES Project, Engineering Town
Jeremy, Bookkeeper of the Dairy Cooperative Tulaga, Engineering Town

Large-Scale Watershed PES in Upper Tana Basin

Fred Kihara, Water Fund Manager, The Nature Conservancy, Nairobi
Charles Lukania Oluchina, Kenya Program Manager
Boniface M. Mwaniki, Regional Manager, WRUA Tana Catchment Area, Embu

P. Aerni, *The Sustainable Provision of Environmental Services*, CSR, Sustainability, Ethics & Governance, DOI 10.1007/978-3-319-19345-8

Joyce Kaburu, Member of WITIKO Women Group (AFS Women Enterprise Loan Fund), Embu

Slaughterhouse Business in Kiserian

Michael Kibue, Communication Officer, Keekonyokie Slaughter House and secretary of the Kenya Livestock Working Group, Kiserian

Kilimo Salama Microinsurance for Farmers

Fritz Brugger, Head of Agricultural Support Services, Syngenta Foundation, Basel
Rose Goslinga, Agricultural Microinsurance Coordinator, Syngenta Foundation, Nairobi
Patrick Karimi Nyaga, Field Coordinator, Syngenta Foundation, Nairobi/Embu
James Nioroge Kiguru, Ass. Agriculture Underwriter, UAP Insurance, Nairobi

Private Standards and Flower Export Business, Thika

Jaime Hernandez, Global Product Stewardship Manager, Syngenta, Basel
Leonard Kipchumba Cherop, Syngenta Sales Representative, Thika
Peter Ochami, Manager of Zena Rose Kenya, Thika
Samuel Kamau, Manager of Kariki, Ltd, Thika
Finn Davey, Chairman WIJBU, Nairobi

Livestock and PES

Sara Namirembe, International Livestock Research Institute (ILRI), Nairobi
Mohamed Said, International Livestock Research Institute (ILRI), Nairobi
Jan de Leeuw, World Agroforestry Center (ICRAF), Nairobi
Pierre Gerber, FAO, Rome
Stephane Forman, Livestock Specialist, The World Bank, Agriculture and Rural Development, Africa Region

References

Abdulai A, CroleRees A (2001) Determinants of income diversification amongst rural households in Southern Mali. Food Policy 26(4):437–452

Aerni P (2005) Stakeholder attitudes towards genetically modified crops in South Africa. Environ Sci Policy 8(5):464–476

Aerni P (2006) The principal-agent problem in international development assistance and its impact on local entrepreneurship in Africa: time for new approaches. ATDF J 3(2):27–33

Aerni P (2007) Exploring the linkages of commerce, higher education and human development: a historical review. ATDF J 4(2):35–48

Aerni P (2009) What is sustainable agriculture? Empirical evidence of diverging views in Switzerland and New Zealand. Ecol Econ 68(6):1872–1882

Aerni P (2011a) Food sovereignty and its discontents. ATDF J 8(1/2):23–40

Aerni P (2011b) Connecting catch-up growth to sustainable development—a new theoretical perspective. ATDF J 8(3/4):29–42

Aerni P (2013) Do private standards encourage or hinder innovation? NCCR trade working paper no. 2013/18, June 2013. World Trade Institute (WTI), University of Bern. http://www.wti.org/fileadmin/user_upload/nccr-trade.ch/wp3/2013_06_final_Private_Standards_Paper_Philipp.pdf

Aerni P, Bernauer T (2006) Stakeholder attitudes toward GMOs in the Philippines, Mexico, and South Africa: the issue of public trust. World Dev 34(3):557–575

Aerni P, Rae A, Lehmann B (2009) Nostalgia versus pragmatism? How attitudes and interests shape the term sustainable agriculture in Switzerland and New Zealand. Food Policy 34 (2):227–235

Aerni P, Häberli C, Karapinar B (2011) Reframing sustainable agriculture. In: Cottier T, Delimatsis P (eds) Prospects of international trade regulation. Cambridge University Press, Cambridge, pp 169–210

Agrivair (2012) 20 years on: 'working together', reviewing experienced and future prospects. http://www.nestle-waters.com/media/featuredstories/Documents/CP%20agrivairUk.pdf

Akerlof GA (1970) The market for" lemons": quality uncertainty and the market mechanism. Q J Econ 84(3):488–500

Ankers P (2001) Imperial ecology. Environmental order in the British empire, 1895–1945. Harvard University Press, Cambridge, MA

Appiah A (2007) What was Africa for them? The New York review of books, 27 September 2007

Arrow K et al (2004) Are we consuming too much? J Econ Perspect 18(3):147–172

Baylis K, Rausser GC, Simon LK (2005) Including non-trade concerns: the environment in EU and US agricultural policy. Int J Agric Resour Gov Ecol 4(3/4):262–276

P. Aerni, *The Sustainable Provision of Environmental Services*, CSR, Sustainability, Ethics & Governance, DOI 10.1007/978-3-319-19345-8

Bayon R (2008) Biodiversity banking: a primer. Ecosystem market place. http://www.ecosystemmarketplace.com/pages/dynamic/article.page.php?page_id=5617&section=home

BBC (2003) Concerns grow over flower farms: BBC one real story, 12 May. news.bbc.co.uk/1/hi/programmes/real_story/3020329.stm

Beck U (1992) Risk society: towards a new modernity. Sage Publications, New York, NY

Beckerman W (2002) A poverty of reason: sustainable development and economic growth. Independent Institute, London

Bennett K (2010) Additionality: the next step for ecosystem service markets. Duke Environ Law Policy Forum 20:417–438

Bennett G, Carroll N (2014) Gaining depth: state of watershed investment 2014. A report by Forest Trends' Ecosystem Marketplace, Washington, DC. Available online at www.ecosystemmarketplace.com/reports/sowi2014

Berndt ER, Glennerster R, Kremer MR, Lee J, Levine R, Weizsacker G, Williams H (2005) Advanced purchase commitments for a malaria vaccine: estimating costs and effectiveness. NBER working paper no. 11288

Blomquist P (2004) The choice revolution: privatization of Swedish welfare services in the 1990s. Soc Policy Adm 38(2):139–155

Boserup E (1976) Environment, population, and technology in primitive societies. Popul Dev Rev 2:21–36

Boserup E (2005) The conditions of agricultural growth: the economics of agrarian change under population pressure. Transaction Publishers, New Jersey, NJ

Buchanan JM, Tullock G (1962) The calculus of consent: logical foundations of constitutional democracy. University of Michigan Press, Ann Arbor, MI

Burgin S (2008) BioBanking: an environmental scientist's view of the role of biodiversity banking offsets in conservation. Biodivers Conserv 17:807–816

Business and Biodiversity Offsets Programme (BBOP) (2013) To no net loss and beyond: an overview of the Business and Biodiversity Offsets Programme (BBOP), Washington, DC. www.forest-trends.org/biodiversityoffsetprogram/guidelines/Overview_II.pdf

Chen X, Viña A, Shortridge A, An L, Liu J (2014) Assessing the effectiveness of payments for ecosystem services: an agent-based modeling approach. Ecol Soc 19(1):7, http://dx.doi.org/10.5751/ES-05578-190107

Chomitz K, Brenes E, Constantino L (1999) Financing environmental services: the Costa Rica experience and its implications. Sci Total Environ 240:157–169

Ciolos D (2011) Getting ready for agricultural research and innovation. Speech by the EU agriculture commissioner in Brussels on March 7, 2011. http://europa.eu/rapid/press-release_SPEECH-12-160_en.htm?locale=en

Clapp J, Fuchs D (2009) Corporate power in the global agrifood governance. MIT Press, Cambridge, MA

Cleaver F (2002) Moral ecological rationality, institutions and the management of common property resources. Dev Change 31(1):361–383

Climate Focus (2012) New market-based mechanisms: options and outlook. Briefing note new market mechanisms, August 2012. http://www.climatefocus.com/documents/files/new_market_mechanism_brief.pdf

Coase RH (1960) The problem of social cost. J Law Econ 3:1–44

Cohen B, Winn MI (2007) Market imperfections, opportunity and sustainable entrepreneurship. J Bus Ventur 22(2007):29–49

Corbera E (2012) Problematizing REDD+ as an experiment in payments for ecosystem services. Curr Opin Environ Sustain 4(6):612–619

Costanza R, d'Arge R, de Groot R, Farber S, Grasso M, Hannon B, Limburg K, Naeem S, O'Neill RV, Paruel J, Raskin RG, Sutton P, van den Belt M (1997) The value of the world's ecosystem services and natural capital. Nature 387:253–260

Cramer C, Johnston D, Mueller B, Oya C, Sender J (2014) How to do (and how not to do) fieldwork on Fair Trade and rural poverty. Can J Dev Stud 35(1):170–185

Daniels AE, Bagstad K, Esposito V, Moulaert A, Rodriguez CM (2010) Understanding the impacts of Costa Rica's PES: are we asking the right questions? Ecol Econ 69(11):2116–2126
Diamond J (2012) The world until yesterday. Viking, New York, NY
Dietz T, Ostrom E, Stern PC (2003) The struggle to govern the commons. Science 301:1907–1912
Ehrlich P, Ehrlich A (2009) The dominant animal: human evolution and the environment. Island Press, Washington, DC
Engel S, Pagiola S, Wunder S (2008) Designing payments for environmental services in theory and practice: an overview of the issues. Ecol Econ 65(4):663–674
European Commission (EC) (2011) Prospects for agricultural markets and income in the EU 2011-2020. DG Agriculture, Brussels
Fairhead J, Leach M, Scoones I (2013) Green grabbing: a new appropriation of nature. Routledge, London
Farinelli F, Bottini M, Akkoyunlu S, Aerni P (2011) Green entrepreneurship: the missing link towards a greener economy. ATDF J 8(3/4):42–48
Farley J, Costanza R (2010) Payments for ecosystem services: from local to global. Ecol Econ 69 (11):2060–2068
Ferraro PJ (2008) Asymmetric information and contract design for payments for environmental services. Ecol Econ 65:811–22
Ferraro PJ, Pattanayak SK (2006) Money for nothing? A call for empirical evaluation of biodiversity conservation investments. PLoS Biol 4(4):e105
Ferraro PJ, Simpson RD (2002) The cost-effectiveness of conservation payments. Land Econ 78 (3):339–353
Ferroni M, Castle P (2011) Public-private partnerships and sustainable agricultural development. Sustainability 3(7):1064–1073
Fisher B, Turner RK, Morling P (2009) Defining and classifying ecosystem services for decision making. Ecol Econ 68:643–653
FONAFIFO, CONAFOR and Ecuador Ministry of Environment (2012) Lessons learned for REDD + from PES and conservation incentive programs. Examples from Costa Rica, Mexico, and Ecuador. http://www.profor.info/sites/profor.info/files/docs/Summary%20of%20PES%20Lessons%20for%20REDD%20March%2019.pdf
Food and Agriculture Organization of the United Nations (FAO) (2007) The state of food and agriculture: paying farmers for environmental services. FAO, Rome
Food and Agriculture Organization of the United Nations (FAO) (2008) Forests and water. FAO forestry paper. FAO, Rome.
Food and Agriculture Organization of the United Nations (FAO) (2010) Grassland carbon sequestration: management, policy and economics. Workshop proceedings. FAO, Rome. http://www.fao.org/docrep/013/i1880e/i1880e00.htm
Food and Agriculture Organization of the United Nations (FAO) (2011a) Payments for ecosystem services and food security. FAO, Rome
Food and Agriculture Organization of the United Nations (FAO) (2011b) Climate change mitigation finance for smallholder agriculture: a guide book to harvesting soil carbon sequestration. FAO, Rome
Food and Agriculture Organization of the United Nations (FAO) (2011c) Building networks for market access. Lessons learned from the Rural Knowledge Network (RKN) pilot project for East Africa (Uganda, Kenya and Tanzania). FAO/Government Cooperative Programme, Rome. http://www.fao.org/docrep/014/i2088e/i2088e00.pdf
Food and Agriculture Organization of the United Nations (FAO) (2012) Decent rural employment for food security: a case for action. FAO, Rome
Freidberg S (2007) Supermarkets and imperial knowledge. Cult Geogr 14:321–342
Freidberg S (2010) Commentary: perspective and power in the ethical foodscape. Environ Plan A 42:1868–1874
Gaupp P (2013) Nespressos Programm für dauerhafte Qualität. Neue Zuercher Zeitung, 15 January 2013

Gichohi WH (2003) Direct payments as a mechanism for conserving important wildlife corridor links between Nairobi National Park and its wider ecosystem: the wildlife conservation lease program. Vth world parks congress, Durban, South Africa, September 2003

Gilbert N (2012) African agriculture: dirt poor. Nature 483:525–527

Gilman N (2004) Mandarins of the future: modernization theory in cold war America. John Hopkins University Press, Baltimore, MD

Goldman RL, Thompson BH, Daily GC (2007) Institutional incentives for managing the landscape: inducing cooperation for the production of ecosystem services. Ecol Econ 64:333–343

Gómez-Baggethun E, Rudolf de Groot R, Lomas PL, Montes C (2010) The history of ecosystem services in economic theory and practice: from early notions to markets and payment schemes. Ecol Econ 69(6):1209–1218

Greenspan Bell R, Russell C (2002) Environmental policy for developing countries. Issues in science and technology, Spring 2002. http://www.issues.org/18.3/greenspan.html

Hanley N, Kirkpatrick H, Simpson I, Oglethorpe D (1998) Principles for the provision of public goods from agriculture: modeling moorland conservation in Scotland. Land Econ 74 (1):102–113

Hardin G (1968) The tragedy of the commons. Science 162:1243–1248

Harman G (1999) Moral philosophy meets social psychology: virtue ethics and the fundamental attribution error. Proc Aristot Soc 99:315–33

Hart SL, Christensen C (2002) The great leap: driving innovation from the base of the pyramid. MIT Sloan Manage Rev 44(1):51–56

Hatanaka M, Bain C, Busch L (2005) Third-party certification in the global agrifood system. Food Policy 30(3):354–369

Hayek F (1988) Fatal conceit. The errors of socialism. Chicago University Press, Chicago

Heal G (1999) New strategies for the provision of global public goods. In: Kaul I, Grunberg I, Stern MA (eds) Global public goods. International cooperation in the 21st century. Oxford University Press, New York, NY

Henson S, Humphrey J (2010) Understanding the complexities of private standards in global agrifood chains as they impact developing countries. J Dev Stud 46(9):1628–1646

Hollander JM (2003) The real environment crisis: why poverty, not affluence, is the environment's number one enemy. University of California Press, Berkeley, CA

IFAD (2011) Stories from the field. International Fund for Agricultural Development, Rome, http://www.ruralpovertyportal.org/documents/654016/690745/DLFE-3795.pdf?version=1.0&t=1337680375350

IFAD (2012) Republic of Kenya Upper Tana Catchment Natural Resource Management Project (UTaNRMP): project design report. East and Southern Africa Division, Project Management Department, Nairobi

International Fund for Agricultural Development (IFAD) (2009) Sustainability of rural development projects. Best practices and lessons learned (the case of Vietnam). IFAD, Rome, Italy, http://www.ifad.org/operations/projects/regions/pi/casestudies/vietnam.pdf

International Livestock Research Institute (ILRI) (2012) Greening livestock: assessing the potential of payment for environmental services in livestock inclusive agricultural production systems in developing countries. ILRI position paper. Nairobi, Kenya. http://cgspace.cgiar.org/bitstream/handle/10568/21188/GreeningLivestock.pdf?sequence=6

Jonas H (1985) Imperative of responsibility: in search of an ethic for the technological age. University of Chicago Press, Chicago

Jones CI, Romer PM (2010) The new Kaldor facts: ideas, institutions, population, and human capital. Am Econ J Macroecon 2(1):224–245

Juma C (2011) Science meets farming. Science 334:1323

Karlen DL, Mausbach MJ, Doran JW, Cline RG, Harris RF, Schuman GE (1997) Soil quality: a concept, definition, and framework for evaluation. Soil Sci Soc Am J 61(1):4–10

Karrer-Rueedi E, Trueb D (2011) Empowering micro-entrepreneurs and small businesses through mobile phones in emerging markets and developing countries. ATDF J 8(3/4):13–18

Kinzig AP, Perrings C, Chapin FS, Polasky S, Smith VK, Tilman D, Turner BL (2011) Paying for ecosystem services—promise and peril. Science 334:603–604

Kleijn D et al (2006) Mixed biodiversity benefits of agri-environment schemes in five European countries. Ecol Lett 9:243–254

Kodhandaraman B, Daniel J (2010) Knowledge transfer for a horticultural revolution: the lifelong learning for farmers model. In: Proceedings of the 28th international horticultural congress, Lisbon, Portugal. http://www.col.org/resources/speeches/2010presentation/Pages/2010-08-22b.aspx

Kosoy N, Corbera E (2010) Payments for ecosystem services as commodity fetishism. Ecol Econ 69:1228–1236

Kraft SE, Lant C, Gillman K (1996) WQIP: an assessment of its chances for acceptance by farmers. J Soil Water Conserv 51:494–498

Kremen C (2005) Managing ecosystem services: what do we need to know about their ecology? Ecol Lett 8:468–479

Laurans Y, Lemenager T, Aoubid S (2012) A savoir: payments for environmental services: from theory to practice—what are the prospects for developing countries? Agence Francaise de Developpement (AFD), Paris

Leach M, Scoones I (2013) Carbon forestry in West Africa: the politics of models, measures and verification processes. Global Environ Change 23(5):957–967

Legrand T, Froger G, Le Coq JF (2013) Institutional performance of payments for environmental services: an analysis of the Costa Rican program. Forest Policy Econ 37:115–123

Leimona B, Joshi L, van Noordwijk M (2009) Can rewards for environmental services benefit the poor? Lessons from Asia. Int J Commons 3(1):82–107

Lele S, Wilhusen P, Brockington D, Seidler R, Bawa K (2010) Beyond exclusion: alternative approaches to biodiversity conservation in developing countries. Curr Opin Environ Sustain 2:94–100

Levy S (2007) Productividad, Crecimiento y Pobreza en México: Qué Sigue Después de PROGRESA-Oportunidades? Inter-American Development Bank, Washington, DC

Li J, Feldman MW, Li S, Daily GC (2012) Rural household income and inequality under the Sloping Land Conversion Program in Western China. Proc Natl Acad Sci U S A 108 (19):7721–7726

Lichtenstein S, Slovic P (2006) The construction of preference. Cambridge University Press, Cambridge

Lockie S (2013) Market instruments, ecosystem services, and property rights: assumptions and conditions for sustained social and ecological benefits. Land Use Policy 31:90–98

Madelev J (2009) Big business, poor peoples: how transnational corporations damage the global poor. Zed Books, London

Madsen B, Carroll N, Moore Brands K (2010) State of biodiversity markets report: offset and compensation programs worldwide. Washington, DC. http://www.ecosystemmarketplace.com/documents/acrobat/sbdmr.pdf

Magnier A, Kalaitzandonakes N, Miller DJ (2010) Product life cycles and innovation in the US seed corn industry. Int Food Agribusiness Manage Rev 13(3), http://ageconsearch.umn.edu/bitstream/93557/2/2.pdf

Malanczuk P (1997) Akehurst's modern introduction to international law. Routledge, London

Mandela N (1995) Long walk to freedom: the autobiography of Nelson Mandela. Back Bay Books, New York, NY

McAfee K (1999) Selling nature to save it? Biodiversity and green developmentalism. Environ Plan D Soc Space 17(2):133–154

McAfee K, Shapiro EN (2010) Payments for ecosystem services in Mexico: nature, neoliberalism, social movements, and the State. Ann Assoc Am Geogr 100(3):579–599

McKenney BA, Kiesecker JM (2010) Policy development for biodiversity offsets: a review of offset frameworks. Environ Manage 45:165–176

McNeill D, Nesheim I, Brouwer F (2012) Land use policies for sustainable development: exploring integrated assessment approaches. Edward Elgar, Cheltenham
Meijerink G (2008) The role of measurement problems and monitoring in PES schemes. In: Dellink RB, Ruijs A (eds) Economics of poverty, environment and natural-resource use. Springer, The Hague
Mekonnen MM, Hoekstra AY, Becht R (2012) Mitigating the water footprint of export cut flowers from the Lake Naivasha Basin, Kenya. Water Resour Manage 26:3725–3742
Millennium Ecosystem Assessment (MEA) (2005) Synthesis report. Island, Washington, DC
Miranda M, Porras IT, Moreno MK (2003) The social impacts of payments for environmental services in Costa Rica: a quantitative field survey and analysis of the Virilla watershed. International Institute for Environmental Development, London
Modelo M (2014) The paradox of fair trade. Stanf Soc Innov Rev Winter:40–45, http://www.ssireview.org/pdf/The_Paradox_of_Fair_Trade.pdf
Moss N, Nussbaum R (2011) A review of three REDD+ safeguard initiatives. UN-REDD programme and the forest carbon partnership facility. http://reddpluspartnership.org/29785-0e1d064b96b75dfb79469cf0d86708dda.pdf
Muller J, Albers HJ (2004) Enforcement, payments, and development projects near protected areas: how the market settings determine what works where. Resour Energy Econ 26:185–204
Muradian R, Corbera E, Pascual U, Kosoy N, May PH (2010) Reconciling theory and practice: an alternative conceptual framework for understanding payments for environmental services. Ecol Econ 69(6):1202–1208
Murphy V (2013) Supermarkets cash in on unfounded fears about food and health. The Guardian, 9 May 2013. http://www.theguardian.com/science/blog/2013/may/09/supermarkets-unfounded-fears-food-health
Najam A, Papa M, Taiyab N (2006) Global environmental governance—a reform agenda. International Institute for Sustainable Development. http://www.iisd.org/pdf/2006/geg.pdf
Namirembe S, Leimona B, van Noordwijk M, Bernard F (2013) Co-investment paradigms as alternatives to payments for tree-based ecosystem services in Africa. Curr Opin Environ Sustain 6:89–97
Nisbet MC (2015) Framing, the media and risk communication in policy debates. In: Cho H, Reimer T, McComas K (eds) Sage handbook of risk communication. Sage, Newbury Park, CA, http://climateshiftproject.org/wp-content/uploads/2014/03/Nisbet_InPress_FramingRiskCommunication_SageHandbookRiskCommunication.pdf
Noleppa S, von Witzke H, Cartsburg M (2013) The social, economic and environmental value of agricultural productivity in the European Union. HFFA working paper 03/2013. http://hffa.info/files/hffa-wp-3-2013.pdf
Norgaard R (2010) Ecosystem services: from eye-opening metaphor to complexity blinder. Ecol Econ 69(6):1219–1227
Nyende P (2011) Lessons learned from the Rural Knowledge Network (RKN) pilot project for East Africa (Uganda, Kenya and Tanzania). FAO/Government Cooperative Programme. FAO, Rome Italy. http://www.fao.org/docrep/014/i2088e/i2088e00.pdf
Oates W, Baumol W (1975) The instruments for environmental policy. In: Mills ES (ed) Economic analysis of environmental problems. National Bureau of Economic Research, Washington, DC, http://www.nber.org/chapters/c2834.pdf
OECD (2001) Sustainable development: critical issues. OECD, Paris
OECD (2011) Regional development policies in OECD countries. OECD, Paris
Ortega JM, Ojea E, Roux C (2013) Payments for water ecosystem services in Latin America: a literature review and conceptual model. Ecosyst Serv 6:122–132
Padulosi S, Bergamini N, Lawrence T (eds) (2012) On farm conservation of neglected and underutilized species: status, trends and novel approaches to cope with climate change. In: Proceedings of an international conference, Frankfurt, 14–16 June 2011. Bioversity International, Rome. http://www.bioversityinternational.org/fileadmin/bioversity/publications/pdfs/1512_On_farm_conservation_of_neglected_and_underutilized_species_status_trends_and_novel_approaches_to_cope_with.pdf

Pagiola S (2007) Guidelines for 'Pro-Poor' payments for environmental services. Environment Department, World Bank, Washington, DC, http://siteresources.worldbank.org/INTEEI/Resources/ProPoorPES-2col.pdf

Pagiola S (2008) Payments for environmental services in Costa Rica. Ecol Econ 65(4):712–724

Pagiola S, Arcenas A (2013) TEEBcase: regional integrated silvopastoral ecosystem management project—Costa Rica, Colombia and Nicaragua. Available at: TEEBweb.org

Pagiola S, Arcenas A, Platais G (2005) Can payments for environmental services help reduce poverty? An exploration of the issues and the evidence to date from Latin America. World Dev 33(2):237–253

Pattanayak SK, Wunder S, Ferraro PJ (2010) Show me the money: do payments supply environmental services in developing countries? Rev Environ Econ Policy 4(2):254–274

Perrot-Maître D (2008) The Vittel payments for ecosystem services: a 'perfect' PES case? International Institute for Environment and Development, London

Pfaff A, Robalino JA, Sanchez-Azofeifa GA (2008) Paying for environmental services: empirical analysis for Costa Rica. Working paper series SAN 08-05. Terry Sanford Institute for Public Policy, Duke University, Durham, NC

Pirard R (2012a) Market-based instruments for biodiversity and ecosystem services: a lexicon. Environ Sci Policy 19–20:59–68

Pirard R (2012b) Payments for Environmental Services (PES) in the public policy landscape: 'Mandatory' spices in the Indonesian recipe. Forest Policy Econ 18:23–29

Pirard R, Billé R, Sembrés T (2010) Upscaling payments for environmental services (PES): critical issues. Trop Conserv Sci 3(3):249–261

Pollan M (2006) Omnivore's dilemma. Bloomsbury, New York, NY

Porras I, Grieg-Gran M, Neves B (2008) All that glitters: a review of payments for watershed services in developing countries. Natural resource issues no. 11. International Institute for Environment and Development (IIED), London, UK

Prager K, Reed M, Scott A (2012) Encouraging collaboration for the provision of ecosystem services at a landscape scale—rethinking agri-environmental payments. Land Use Policy 29 (1):244–249

Prins G, Rayner S (2007) Time to ditch Kyoto. Nature 449:973–975

CH Bundesnetzwerk laendlicher Raum (2012) Modellvorhaben «Synergien im ländlichen Raum» Anreize zu mehr Innovation und verstärkter sektorübergreifender Zusammenarbeit. Bundesamt fuer Raumplanung, Bern, Switzerland. http://www.are.admin.ch/dokumentation/publikationen/00022/00505/index.html?lang=de

Rawlings LB, Rubio GM (2005) Evaluating the impact of conditional cash transfer programs. World Bank Res Obs 20(1):29–55

Reardon S (2012) China leads March for green economy. Special report. New Scientist 16 June:8–9

Renting H, Rossing WAH, Groot JCJ, Van der Ploeg JD, Laurent C, Perraud D, Stobbelaar DJ, Van Ittersum MK (2009) Exploring multifunctional agriculture. A review of conceptual approaches and prospects for an integrative transitional framework. J Environ Manage 90 (2):112–123

Riisgaard L (2007) What's in it for labour? Private social standards in the cut flower industries of Kenya and Tanzania. Diis working paper 2007/16. Danish Institute of Development Studies, Copenhagen

Rodrik D (2013) The past, present, and future of economic growth. Working paper no. 1. Global Citizen Foundation. http://www.gcf.ch/?page_id=5758

Romer PM (1990) Endogenous technological change. J Polit Econ 98:71–102

Romer PM (1994) New goods, old theory, and the welfare costs of trade restrictions. J Dev Econ 43:5–38

Sabau GL (2010) Know, live and let live: towards a redefinition of the knowledge-based economy—sustainable development nexus. Ecol Econ 69(6):1193–1201

Schiermeier Q (2012) The Kyoto protocol: hot air. Nature 491:656–658

Schumpeter J (2008) Capitalism, socialism and democracy (first published in 1942). Harper Perennial, New York, NY

Shell D, Meijaard E (2010) Purity and prejudice: deluding ourselves about biodiversity conservation. Biotropica 42(5):566–568

Shogren JF, Parkhurst GM, Settle C (2003) Integrating economics and ecology to protect nature on private lands: models, methods, and mindsets. Environ Sci Policy 6:233–242

Sills E, Arriagada RA, Ferraro PJ, Pattanayak SK, Carrasco LE, Ortiz E, Cordero S, Caldwell K, Andam K (2008) Private provision of public goods: evaluating payments for eco system services in Costa Rica. Working paper. North Carolina State University, Raleigh, VA

Silvestri S, Osano P, de Leeuw J, Herrero M, Ericksen P, Kariuki J, Njuki J, Bedelian C, Notenbaert A (2012) Greening livestock: assessing the potential of payment for environmental services in livestock inclusive agricultural production systems in developing countries. ILRI, Nairobi, http://cgspace.cgiar.org/bitstream/handle/10568/21188/GreeningLivestock.pdf?sequence=6

Stavins RN (2005) Economics of the environment. W.W. Norton, New York, NY

Swift MJ, Izac A-MN, van Noordwijk M (2004) Biodiversity and ecosystem services in agricultural landscapes—are we asking the right questions? Agric Ecosyst Environ 104(1):113–134

TEEB (2009) The economics of ecosystems and biodiversity for national and international policy makers. UNEP-TEEB Office, Geneva, http://www.teebweb.org/teeb-study-and-reports/main-reports/national-and-international-policy-making/

The Katoomba Group (TKG) (2009) Beyond carbon. Biodiversity and water markets. http://ecosystemmarketplace.com/documents/acrobat/ECM%20Beyond%20Carbon_eng.pdf

Tinbergen J (1952) On the theory of economic policy. North Holland, Amsterdam

Tran P, Faust M, Shaw R (2010) Catastrophic flood and forest cover change in the Huong river basin, central Viet Nam: a gap between common perceptions and facts. J Environ Manage 91 (11):2186–2200

Turner RK, Pearce D, Bateman I (1994) Environmental economics: an elementary introduction. John Hopkins University Press, Baltimore, MD

Turner MG, Gardner RH, O'Neill RV (2001) Landscape ecology in theory and practice—pattern and process. Springer, New York, NY

UN Division for Sustainable Development (UN-DESA), United Nations Environment Programme (UNEP) and UN Conference on Trade and Development (UNCTAD) (2012) The transition to a green economy: benefits, challenges and risks from a sustainable development perspective. Second preparatory committee meeting for United Nations Conference on Sustainable Development (UNCSD), Rio, Brazil. http://www.unep.org/greeneconomy/Portals/88/documents/research_products/UN-DESA,%20UNCTAD%20Transition%20GE.pdf

United Nations (2012) The future we want: Rio+20 outcome statement (A/CONF.216/L.1). Rio de Janeiro, Brazil, 20–22 June 2012. http://www.stakeholderforum.org/fileadmin/files/FWWEnglish.pdf

Van Hecken G, Bastiaensen J (2010) Payments for ecosystem services: justified or not? A political view. Environ Sci Policy 13(8):785–792

Van Noordwijk M et al (2012) Payments for environmental services: evolution toward efficient and fair incentives for multifunctional landscapes. Annu Rev Environ Resour 37:389–420

Vatn A (2010) An institutional analysis of payments for environmental services. Ecol Econ 69 (6):1245–1252

Warsh D (2006) Knowledge and the wealth of nations: a story of economic discovery. W. W. Norton, New York, NY

Wen C, Siu YL, Hubacek K (2012) Carrot and stick—a novel policy experiment of transboundary watershed protection in China. Environ Sci Technol 46:6451–6452

Willy DK, Kuhn A, Holm-Mueller K (2012) Payments for environmental services (PES) and the characteristics of social ecological systems: the Case of Lake Naivasha Basin. Institute for Food and Resource Economics University of Bonn. Agricultural and resource economics discussion paper 2012:5. http://www.ilr.uni-bonn.de/agpo/publ/dispap/download/dispap12_05.pdf

Winters P, Chiodi V (2010) Human capital investment and long-term poverty reduction in rural Mexico. J Int Dev 23(4):515–538

Winters P, Davis B (2009) Designing a programme to support smallholder agriculture in Mexico: lessons from PROCAMPO and Oportunidades. Dev Policy Rev 27(5):617–642

Wongtschowski M, Belt J, Heemskerk W, Kahan D (2013) The business of agricultural business services—working with smallholders in Africa. Food and Agriculture Organization of the United Nations (FAO), Rome

World Bank (2012) State and trends of the carbon market in 2012. Washington, DC. http://siteresources.worldbank.org/INTCARBONFINANCE/Resources/State_and_Trends_2012_Web_Optimized_19035_Cvr&Txt_LR.pdf

World Resources Institute (WRI) (2008) Ecosystem services: a guide for decision makers. Island Press, Washington, DC

Worldwatch Institute (2008) State of the world 2008. Innovations for a sustainable economy. WRI, Washington, DC, http://www.worldwatch.org/files/pdf/SOW08_chapter_9.pdf

Wunder S (2005) Payments for environmental services: some nuts and bolts. CIFOR occasional paper no. 42. Center for International Forestry Research, Bogor, Indonesia

Wunder S (2008) Necessary conditions for ecosystem service payments. Economics and conservation in the tropics: a strategic dialogue, January 31–February 1. http://www.rff.org/documents/08_tropics_conference/tropics_conference_papers/tropics_conference_wunder_pes_markets.pdf

Wunder S, Boerner J (2011) Changing land uses in forestry and agriculture through payments for environmental services. In: Hung-Hong Y et al (eds) The environment, climate change, and land policies. Lincoln Institute, Cambridge, MA

Wunder S, Engel S, Pagiola S (2008) Taking stock: a comparative analysis of payments for environmental services programs in developed and developing countries. Ecol Econ 65:834–852

Zilberman D, Lipper L, McCarty N (2008) When could payments for environmental services benefit the poor? Environ Dev Econ 13:255–278

Index

P. Aerni, *The Sustainable Provision of Environmental Services*, CSR, Sustainability, Ethics & Governance, DOI 10.1007/978-3-319-19345-8

MIX
Papier aus verantwortungsvollen Quellen
Paper from responsible sources
FSC® C105338

If you have any concerns about our products,
you can contact us on
ProductSafety@springernature.com

In case Publisher is established outside the EU,
the EU authorized representative is:
Springer Nature Customer Service Center GmbH
Europaplatz 3, 69115 Heidelberg, Germany

Printed by Libri Plureos GmbH
in Hamburg, Germany